To Mario Geymonat

Michele Emmer *Editor*

Imagine Math

Between Culture and Mathematics

 Springer

Michele Emmer
Department of Mathematics
Sapienza University of Rome

ISBN 978-88-470-5827-9   ISBN 978-88-470-2427-4 (eBook)
DOI 10.1007/978-88-470-2427-4

Springer Milan Dordrecht Heidelberg London New York

© Springer-Verlag Italia 2012
Softcover re-print of the Hardcover 1st edition 2012

Translation: Kim Williams for the contributions by E. Fabbri, E. Gamba, M. Costamagna, S. Carandini, C. Viviani and M. Emmer ("Visual Harmonies: an Exhibition on Art and Math" and "The Fantastic World of Tor Bled-Nam")

Cover-Design: deblik, Berlin
Cover-Image: Lucio Saffaro, "Il piano di Orfeo" (particular). Courtesy of *Fondazione Lucio Saffaro*, Bologna; Springer thanks Gisella Vismara
Typesetting with LaTeX: CompoMat S.r.l., Configni (RI)

Springer-Verlag Italia S.r.l., Via Decembrio 28, I-20137 Milano
Springer fa parte di Springer Science + Business Media (www.springer.com)

Contents

Introduction

Imagine all the people
Sharing all the world ...
John Lennon

Imagine mathematics, imagine with the help of mathematics, imagine new worlds, new geometries, new forms. Imagine building mathematical models that make it possible to manage our world better, imagine solving great problems, imagine new problems never before thought of, imagine music, art, poetry, literature, architecture, theatre and cinema with mathematics. Imagine the unpredictable and sometimes irrational applications of mathematics in all areas of human endeavour.

Imagination and mathematics, imagination and culture, culture and mathematics. For some years now the world of mathematics has penetrated deeply into human culture, perhaps more deeply than ever before, even more than in the Renaissance. In theatre, stories of mathematicians are staged; in cinema Oscars are won by films about mathematicians; all over the world museums and science centres dedicated to mathematics are multiplying. Journals have been founded for relationships between mathematics and contemporary art and architecture. Exhibitions are mounted to present mathematics, art and mathematics, and images related to the world of mathematics.

"Imagine Math" is intended to contribute to grasping how much that is interesting and new is happening in the relationships between mathematics, imagination and culture.

A look at the past, at figures and events that made history, can also help to understand the phenomena of today.

It is no coincidence that this volume contains an homage to the great Italian artist of the 1700s, Andrea Pozzo, and his perspective views. It also has a particular focus on Luca Pacioli, who remains today a source of inspiration not only for art, but for theatre.

Here theatre, art and architecture are the topics of choice, along with music, literature and cinema.

No less important are applications of mathematics to medicine and economics. Nor does this particular focus neglect the universality of mathematics, from its origins in Mesopotamia up to the geometry of Japanese origami.

The topics are treated in a way that is rigorous but captivating, detailed but full of evocations. An all-embracing look at the world of mathematics and culture.

Michele Emmer

Emmer M. (Ed.): Imagine Math. Between Culture and Mathematics
DOI 10.1007/978-88-470-2427-4_1, © Springer-Verlag Italia 2012

Homage to Benoît Mandelbrot

The Fantastic World of Tor' Bled-Nam

Michele Emmer

Let us imagine that we have been travelling on a great journey to some far-off world. We shall call this world Tor' Bled-Nam. Our remote sensing device has picked up a signal which is now displayed on a screen in front of us. The image comes into focus and we see. [...] What can it be? Can it be some strange looking insect? Or could it be some vast and oddly shaped alien city, with roads going off in various directions to small towns and villages nearby? Maybe it is an island – and then let us try to find whether there is a nearby continent with which it is associated. This we can do by 'backing away', reducing the magnification of our sensing device by a linear factor of about fifteen. Lo and behold, the entire world springs into view. [...] We may explore this extraordinary world of Tor' Bled-Nam as long as we wish, tuning our sensing device to higher and higher degrees of magnification. We find an endless variety: no two regions are precisely alike – yet there is a general flavour that we soon become accustomed to. [...] What is this strange, varied and most wonderfully intricate land that we have stumbled upon? No doubt many readers will already know. But some will not. This world is nothing but a piece of abstract mathematics – the set known as the Mandelbrot set [1, p. 74-79].

The journey into the land of Tor' Bled-Nam begins the chapter that Roger Penrose dedicates to the relationship between mathematics and reality in the book *The Emperor's New Mind*. It may seem paradoxical that the Mandelbrot set is the first example that Penrose uses to confirm the Platonic reality of mathematical concepts, objects that can only be seen on the monitor of a computer! But for Penrose the Mandelbrot set is an astounding example of how human thought can be guided towards an eternal truth that has it own reality and that is only partially revealed to some of us.

Its wonderfully elaborate structure was not the invention of any one person, nor was it the design of a team of mathematicians. Benoit Mandelbrot himself, the Polish-American mathematician (and protagonist of fractal theory) who first studied the set, had no real prior conception of the fantastic elaboration inherent in it, although he knew that he was on the track of something very interesting. Indeed, when his first computer pictures began to emerge, he was under the impression that the fuzzy structures that he was seeing were the result of a computer malfunction (Mandelbrot 1986)! Only later did he become convinced that they were really there in the set itself [1, p. 95].

Michele Emmer
Department of Mathematics, Sapienza University of Rome (Italy).

Emmer M. (Ed.): Imagine Math. Between Culture and Mathematics
DOI 10.1007/978-88-470-2427-4_2, © Springer-Verlag Italia 2012

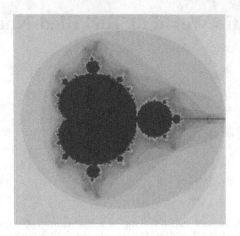

Fig. 1 Mandelbrot set

Penrose adds:

> The complete details of the complication of the structure of Mandelbrot's set cannot really be fully comprehended by any one of us, nor can it be fully revealed by any computer. It would seem that this structure is not just part of our minds, but it has a reality of its own [...] The computer has been used in essentially the same way that the experimental physicist uses a piece of experimental apparatus to explore the structure of the physical world. [...] The Mandelbrot set is not an invention of the human mind. It was a discovery: like Mount Everest, the Mandelbrot set is just *there*! [1, p. 95].

We may not know whether it was a creation or a discovery, but we do know who the inventor/discoverer of the Mandelbrot set and the theory of fractals was: Benoît Mandelbrot, who coined the word *fractal* in 1975. A paragraph from his *Les objets fractals. Forme, hasard et dimension* [2] describes the origins of the term:

> This etymology can be asserted with full authority because I am responsible for coining the term to denote a collection of concepts and techniques that seems finally to acquire a clear-cut identity. *Fractal* comes from the Latin adjective *fractus*, which has the same root as *fraction* and *fragment* and means "irregular or fragmented"; it is related to *frangere*, which means "to break" [3, p. 4].

Mandelbrot, then a mathematician working for IBM, sought to characterise a class of objects in sectors other than mathematics and in some of its applications.

Looking back in 1984 to his first experiences with fractal geometry, Mandelbrot wondered:

> Why is geometry often described as cold and dry? One reason lies in its inability to describe the shape of a cloud, a mountain, a coastline, or a tree. Clouds are not spheres, mountains are not cones, coastlines are not circles, and bark is not smooth, nor does lightning travel in straight line. [...] Nature exhibits not simply a higher degree but an altogether different level of complexity [4, p. v].

Fractal geometry immediately presented itself as the geometry best suited to study the complexity of natural forms and their evolution. Mandelbrot wrote:

Today, however, there is more to geometry than Euclid. During the 1970s it was my privilege
to conceive and develop fractal geometry, a body of thoughts, formulas and pictures that can
be called either a new *geometry of nature* or a new *geometric language* [7, p. 11].

In a 1990 article in *Scientific American* [5] several of the authors of the most evocative fractal images reiterated the fact that fractal geometry seemed to describe the
forms and configurations of nature in a way that was not only succinct but also
aesthetically more valid than traditional Euclidean geometry. The article underlined
the correspondence between fractals and the modern theory of chaos as a sign of a
profound relationship: 'Fractal geometry is the geometry of chaos'.

It further stated that fractals themselves were above all the language of geometry.
This explicit attempt to posit the geometry of fractals as the only true geometry of
nature, the key that made it possible to read and comprehend natural phenomena, is
one of the reasons that they came to be view with suspicion by some scientists. The
history of the idea of space makes it evident that the idea of a universal geometry is
completely illusory and erroneous. Every geometry, every conception of space, can
be useful according to the problem viewed.

The principal property of fractal objects are those of being self-similar and of
having a fractional or fractal dimension, that is, one that is not an integer. A point
has zero dimensions; a line segment or curve has one dimension; a surface has two
dimensions, and so forth. Mathematicians have discovered that there exist mathematical objects that have a dimension comprised, for example, between one and
two: a normal curve has one dimension, it is a line; but if this curve winds on itself
and has many indentations, or jags – like the edge of a part of the coastline of Great
Britain – then it has a dimension that is greater than one, even through it remains a
line and thus has a dimension less than two.

A shape is called self-similar if it can be subdivided into a large number of parts,
each of which is an exact replica at reduced scale of the original. In reality, self-
similarity should be thought of more as an approximate property than an exact one.

Examples of shapes with this property have been known for more than a century, such as the snowflake curve, which is the boundary (the outline) of the shape.
The fractal dimension is another property that generates the infinite jaggedness that
characterise fractals. For example, the fractal dimension of the snowflake curve is
$\log 4/\log 3 = 1.26$.

The great complexity of the images obtained might give the idea that it is very
complicated to understand what the Mandelbrot set is, how it can be defined, how
we can enter the land of Tor' Bled-Nam. But the definition of the Mandelbrot set
is surprisingly simple. In addition to the main cardoid, this set consists in a series
of filaments on which appear miniscule copies of the central figure, copies that are
visible only when the figure is enlarged. To describe the set mathematically, to each
point on the plane is assigned a pair of numbers (a, b), which are the Cartesian
coordinates relative to the vertical and horizontal axes; this is equivalent to assigning
a complex number $z = a + ib$, where i is such that $i^2 = -1$. We then consider the
operation that consists in substituting, for the initial point z, the point $z^2 + c$, where
c is a fixed complex number.

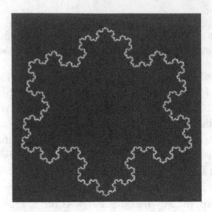

Fig. 2 The snowflake curve

The new value $z^2 + c$, for each point z, represents a new point in the plane. If for example, we initially choose $z = c = 1$ (that is, the complex number $1 + i0 = 1$), through iterations of the operation we obtain:

$$1^2 + 1 = 2, 2^2 + 1 = 5, 5^2 + 1 = 26, ...;$$

if instead $c = -1$, we obtain:

$$(-1)^2 - 1 = 0, 0 - 1 = -1, (-1)^2 - 1 = 0, ...$$

Thus there are only two possibilities: in the first case, the numbers contained in the set assume values that are increasingly higher, and the set is unbounded; in the second case, the set of values is bounded. The Mandelbrot set is the set of points (a, b) for which the set of numbers obtained by iterations of the operation is bounded.

The boundary of the Mandelbrot set is the fundamental example of a fractal. In the introduction to the exhibition entitled *Frontiers of Chaos*, which travelled around the world in the late 1980s, Robert Osserman observed:

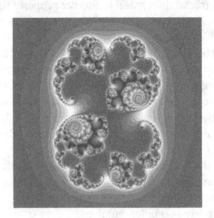

Fig. 3 The Julia set

Mandelbrot's discovery of the figure that carries his name did not come about through simply playing with numbers or formulae to see what would happen. Rather, it was achieved
as a natural consequence of the study of a very ancient problem: the attempt to understand
the complex structure found every time a single operation like $z^2 + c$ is performed over and
over. In these cases we speak of iterative process or iteration. An enormous advantage of
iterative processes is that they are particularly suited to run on a computer. A slight variation
of the procedure that defines the Mandelbrot set using the iteration of the same expression
$z^2 + c$ produces a manifold of a completely different shape. The boundaries are still fractals
called the Julia set, after the mathematician Gaston Julia who studied and introduced them
in 1918. Many physical systems, such as planetary orbits, can be analysed by means of an
iterative process. The mathematical theory that derives from such processes is known by
the name of dynamical systems. Numerous images in this exhibition are obtained through
the application of the methods of dynamical systems to a particular model of magnetism. It
is becoming increasingly clear that shapes with fractal behaviour are present in many situations in the natural world. It may in any case come as a surprise that study of the physics
of magnetism led necessarily to shapes that were almost identical to the Mandelbrot set, a
purely mathematical creation [6].

Since fractal geometry has produced so many new images, those who have created
them could not help but invade art's territory. In compiling their book *The Beauty
of Fractals* [4], Heinz-Otto Peitgen and Peter Richter wanted not only to present
the mathematical theory, but to use mathematical ideas as illustrations of – if not as
an out-and-out pretext for – their creative activity, not so much as mathematicians,
but as artists. In the book's introduction they explicitly refer to the possibility of a
rejoining between scientific language and artistic language:

> *Science and Art*: Two complementary ways of experiencing the natural world – the one
> analytic, the other intuitive. We have become accustomed to seeing them as opposite poles,
> yet don't they depend on one another? The thinker, trying to penetrate natural phenomena
> with his understanding, seeking to reduce all complexity to a few fundamental laws – isn't
> he also the dreamer plunging himself into the richness of forms and seeing himself as part
> of the eternal play of natural events? [4, p. 1].

In his invited contribution to Peitgen and Richter's book, entitled *Fractals and the
Rebirth of Iteration Theory*, Mandelbrot added:

> On its contribution to science and aesthetics, the conclusion is that there was not even an
> inkling of fractal geometry before my work [4, p. 159].

Mandelbrot himself has reiterated the importance of fractals in art:

> Fractal geometry appears to have created a new category of art, next to art for art's sake and
> art for the sake of commerce: art for the sake of science (and of mathematics). [...] The
> source of fractal art resides in the recognition that very simple mathematical formulas that
> seem completely barren may in fact be pregnant, so to speak, with an enormous amount of
> graphic structure. The artist's taste can only affect the selection of formulas to be rendered,
> the cropping and the rendering. Thus, fractal art seems to fall outside the usual categories
> of 'invention', 'discovery' and 'creativity' [7, p. 11, 14].

Elsewhere he states this in even more explicit terms:

> Today we can say that the abstract beauty of the theory is flanked by the plastic beauty of
> the curve, a beauty that is astounding. Thus, within this mathematics that is a hundred years
> old, very elegant from a formal point of view, very beautiful for specialists, there is also a

physical beauty that is accessible to everyone. [...] By letting the eye and the hand intervene in the mathematics, not only have we found again the ancient beauty, which remains intact, but we have also discovered a new beauty, hidden and extraordinary. [...] Those who are only concerned with practical applications may perhaps tend not to insist too much on the artistic aspect, because they prefer to entrench themselves in the technicalities that appertain to practical applications. But why should the rigorous mathematician be afraid of beauty? [8, pp. 18, 19, 29].

Fig. 4 Benoît Mandelbrot

Benoît Mandelbrot was born in Warsaw on 20 November 1924, and died in Cambridge, Massachusetts on 14 October 2010. In 1936 his family moved to Paris, where he began to study mathematics, where his uncle Szolem Mandelbrot was a well-known mathematician at the Collége de France. In 1993 he was awarded the prestigious Wolf Foundation Prize in Physics, for having transformed our vision of nature. The obituary published on 16 October 2010 in *The New York Times* pointed out that Mandelbrot had contributed to research in geology, medicine, cosmology and engineering, using the theory of fractals to explain the clusters of the galaxy and analyse economic theories, frequently in the face of scepticism on the part of the specialists in the various fields. In Italy, perhaps I missed other articles, brief tributes, written by those who cared about the death of a mathematician famous the world over. As we know, the Italians have plenty of other things to fill the pages of their newspapers. The most interesting obituary of Mandelbrot was that written by his cousin, the physicist Jacques Mandelbrot, which appeared on the website of

Leonardo, the most important journal for art, science and technology, published by MIT press:

> He was an extremely original scientist who with the invention of fractals created a new concept which has applications in numerous fields of science and art. His unconventional approach was well accepted when he came to IBM. He was also Professor at Yale University. [...] Concerning art, René Huyghe in his book *Formes et Forces* [10] (*Shapes and Forces*) makes a distinction between art based on shapes, actually shapes which can be described by Euclidian geometry such as are encountered in Classical art, and art like Baroque art based on the action of forces, for instance shapes which are encountered in waves, in tourbillions etc. We could now assert that both Classical and Baroque art can be described geometrically, the first one by Euclidian geometry, the second one by fractal geometry. In sciences, ranging from physics at all scales to economics; fractals give new insights and give a suitable framework to chaos or to phenomena which were outside the mainstream of science due to their complexity. Concerning technology Benoit Mandelbrot was the first one to be surprised when he saw weird and complex shapes appear on his computer screen resulting from an equation. This was to lead to the Mandelbrot set and is the origin of fractal art, a main branch of computer art. Fractals can subconsciously suggest that each of us is a microcosm, an image of the whole world, hence their strong appeal [9].

In closing, I would like to highlight the last words of this tribute, and suggest that the universal aspect of fractals might correspond to the fact that I can subconsciously imply that the small part of the universe that is us, is an image of the entire universe; in other words, that we are a microcosm. I think that Benoît Mandelbrot would be pleased by this phrase of his cousin's.

References

1. R. Penrose, The Emperor's New Mind: Concerning Computers, Minds and The Laws of Physics. Oxford University Press, Oxford, 1989.
2. B.B. Mandelbrot, Les objets fractals. Forme, hasard et dimension. Flammarion, Paris, 1975.
3. B.B. Mandelbrot, Fractals: Forms, Chance and Dimension. Freeman, San Francisco, 1977.
4. H.-O. Peitgen, P.H. Richter, The Beauty of Fractals, Images of Complex Dynamical Systems. Springer, Berlin, 1986.
5. H. Jürgens, H.-O. Peitgen, D. Saupe, The Language of Fractals. Scientific American 263: 60-67, 1990.
6. R. Osserman, I frattali: le frontiere del caos. In: M. Calvesi, M. Emmer (eds.), I frattali: la geometria dell'irregolare. Istituto Enciclopedia Italiana, Roma, pp. 71-83, 1988.
7. B.B. Mandelbrot, Fractals and an Art for the Sake of Science. In: M. Emmer (ed.), The Visual Mind. MIT Press, Cambridge MA, pp. 11-14, 1993.
8. B.B. Mandelbrot, La geometria della natura. Theoria, Roma, 1989.
9. J. Mandelbrot, Fractals in Art, Science and Technology. Leonardo Almanac Online. http://www.leoalmanac.org/index.php//lea/entry/fractals_in_art_science_and_technology. Last accessed 19 July 2011.
10. R. Huyghe, Formes et Forces. Flammarion, Paris, 1971.

Note: If you haven't already understood, try reading Tor' Bled-Nam backwards.

Venice and "La Fenice"

The Reconstruction of the *Teatro La Fenice*: "splendidezze" and "dorature" (Gleam and Gilding)

Elisabetta Fabbri

Nomen est omen, warned Plautus.

Destroyed by fire in 1996, Venice's theatre *La Fenice*, The Phoenix, true to its name, was restored to its original state, and upon its inauguration in December 2003 it received a mostly positive welcome.[1]

The apparently most logical hypothesis – that the theatre building would be re-created as a modern structure – was instead overruled by a desire for conservation on the one hand, and what was certainly an emotional impulse on the other.

It should be recalled that the entire city (or better, the unanimous voice of the city council representing the city) called for the immediate reconstruction of *its* theatre, but it is also true that in its urban dimension the theatre building had not been completely destroyed: the imposing perimeter walls had contained the flames within the building's interior, and the façade as well as other external elements remained intact. Thus the reconstruction had to take into account the conservation of what remained, as Paolo Morachiello and Mario Piana [1] had immediately made clear.

Everything that had survived – the fragments of masonry, sections of surfaces, single decorative elements – became crucial references for the reconstruction. Even where the extent of the reconstructed component was larger than what had been salvaged, the new took its origins and meaning from the old, which, as Aldo Rossi said in his account of the project, *guarantee the historic continuity of the reborn La Fenice*. For this reason it was important to leave the marks of time visible, while stitching the fragments together with the aim of obtaining halls that are *pleasingly restored*.

The *impossible* challenge was that of rebuilding the *interior* of the theatre in order to restore the image of what had been lost. *Which* La Fenice was to be rebuilt was indicated in the project of Aldo Rossi, who proposed reconstructing the theatre auditorium designed in the mid-nineteenth century by the Meduna brothers. This was made possible thanks to *photographic* archives which provided visible images

Elisabetta Fabbri
Architect, Venice (Italy).

[1] The design was created by architect Aldo Rossi. After his death in 1997, it was developed and carried out by the architects Marco Brandolisio, Giovanni Da Pozzo, Massimo Scheurer and Michele Tadini, and engineer Edoardo Guenzani, with the contribution of engineer Nicola Berlucchi for the restoration of the decorative apparatus.

Emmer M. (Ed.): Imagine Math. Between Culture and Mathematics
DOI 10.1007/978-88-470-2427-4_3, © Springer-Verlag Italia 2012

of what had been lost, and to documentary testimony describing the competition won by the Meduna brothers, who were architects and set designers.

In 1853 the competition for a new decorative design for the auditorium was announced. The competition guidelines clearly defined the decorative effect that was to be achieved, asking that the decoration of La Fenice be *gleaming with ornaments and gilding* against *light coloured backgrounds*. The theatre was be resonant and grandiose [2].

Meduna wrote that he had sought to maintain an effect of *elegance conjoined to lavishness*, to make it so that *some parts would be less ostentatious*, and that '*the splendour of its striking beauty* would be apparent everywhere.

News reports of the day describe how the spectator could *feast on the rich splendours of the 1500s, 1600s and 1700s*, and how *lavish elegance* was evident everywhere, since there was *every most posh and sumptuous thing imaginable*, concluding that *whoever sees it swears that the magnificent decorations at Versailles are no greater than these*. Meduna's La Fenice was enthusiastically embraced by the theatre-going public of the day, who were awed by the *luxurious style of the 1600s, which is now most in vogue* [3].

All of these concepts, the transformation of these *statements* into forms, became detailed *illustration* in the decorative system of the auditorium redesigned by the set designer Mauro Carosi, who retraced the nineteenth-century design process, finally arriving at the rediscovery and reinterpretation of the antique spirit of the artistic decoration of the auditorium.

The discovery within the decorative traces of the so-called *teatri di verzura*, theatres made entirely of greenery, which were popular in the 1500s and with which Meduna was certainly familiar, provided the basis for the recreation of those same themes, and this underlying theme is what ties together the whole decor. Decoration is code and language; it is theme and argument. The transformation of garlands, leaves and flowers into faces and animals is a game of anamorphosis. Is it a leaf that is transformed into a face, or a face that is transformed into a leaf? It is a garland that is transformed into a swan. A *bucolic kingdom* already existed in the lost Fenice, and perhaps for a long time no one had been able to recognise and *see* it.

Meduna had told a story that Mauro Carosi rediscovered: entering into the auditorium the face of a satyr, *a man of the forest*, with a beard of leaves and limbs transformed into garlands of acanthus leaves, opens his arms wide in a symbolic embrace of the spectators, marking the entrance into a magic forest, an archaic world that belongs to an age when nature was still sacred.

Hidden among the acanthus leaves that surmount and envelop all of the architecture are heads of children, men, women disguised as animals; there are the *phantoms of the forest*, putti, nymphs, genies, swans and griffons; there are all the characters who *know* how to live in an enchanted forest. The spectators, opening the door to the boxes or entering the auditorium, step into *this forest*, participate in this representation within a representation, becoming part of the *game*.

The redesign of the decorative theme, the interpretation of the *teatro di verzura* created by Carosi, is conjoined to the theme of the geometric reconstruction of the decorative system. Architecture and decoration are tied and held together by the geometry of the hall; even though it is true that the architecture practically *disappears* in the lush decoration, an exact understanding of the geometry was indispensable for the application of the ornamentation.

Set painter Fabio Mattei reconstructed a 3D model of the auditorium vault on the computer, and then transformed the virtual model into an actual wooden model at a scale of 1:10. In order to be able to meet the deadlines imposed, it was necessary to anticipate the design and execution of the vault while there was still no sign of the vault on the construction site. Mattei wrote, *You had to imagine. Imagine that concave ceiling, find the system for measuring the curved surfaces, create a virtual ceiling in the studio that would anticipate the shape of the real one.* This made it possible to perform exact checks of the decorative modules of the ceiling, and to invent, or better, to devise a technique that was capable of exactly reproducing the decoration on the actual vault. Scientific knowledge merged with artistic knowledge to make possible the reconstruction off-site of the complicated play of ornamentation, which was then *mounted* on the empty skeleton of the auditorium in only seven months, thanks to the precise programming of the job and the careful supervision of the construction managers. The strong point of the reconstruction lies in the method used and attention to the detail.

In order to reconstruct La Fenice it was necessary to reclaim the *know-how* of a by-gone age, to reconstruct not only what can be seen, but also to reconstruct *how it is done* by studying an older, knowing tradition of working: *the work of eyes and hands and memory*, as Baricco wrote, *is knowledge saved from oblivion* [4], that of the artisan who puts his skills to the test to recreate a tradition that is lost and no longer in use.

The reconstructed Fenice is an example of the application of a rigorous methodology to give a new form to the theatre that was lost. When visiting the auditorium, you have to know how to read all of that *splendour of ornament and gilt* that was so beloved by Giovan Battista Meduna, and rediscover the bucolic kingdom that was recreated by Mauro Carosi.

In evaluating the final results, it is also necessary to *keep an open mind*, and try to understand what the Teatro La Fenice *really* was before the fire.

The decorative richness that has been reconstructed may be *pleasing* to many, and *horrible* to others, but La Fenice has never been a perfect theatre. As soon as the theatre was built by Selva in 1792, a satirical poem was composed about it:

> *belle pietre, bei legnami*
>
> *scale larghe, palchi infami.*
>
> (handsome stones, handsome wood,
>
> wide stairs, boxes no good)

Similarly, after the first fire and the reconstruction of 1851-57 following the design of the Meduna brothers, mentioned earlier, Pietro Selvatico described La Fenice as *a jumble of ostentatious Baroque ornament superimposed on rigid, classical line.* Be that as it may, with its indubitable wealth and abundance of ornament, it represented and still represents a part of the living *history* of the city, and this is how it has been reconstructed.

Nevertheless, both the reconstruction of the auditorium and the restoration / reconstruction of the Apollonean Halls, with the operative methodologies used to stitch the fragments of decoration together and the solutions adopted, with necessary compromises imposed by realities of the construction site as the work went forward, make it evident that the Teatro La Fenice – with its halls restored and *recomposed* and its auditorium reconstructed – is not an *identical* copy, but an *evocation of the splendour of the past*.

Fig. 1 Foyer of the theatre, right side, after the fire

Fig. 2 Foyer of the theatre, right side, after restoration (courtesy of Teatro La Fenice, Venice)

Fig. 3 Sala Grande after the fire

Fig. 4 Sala Grande after restoration (courtesy of Teatro La Fenice, Venice)

Fig. 5 Theatre auditorium: zenithal view after the fire

Fig. 6 Theatre auditorium, June 2003: restored wooden structure of the boxes

Fig. 7 Theatre auditorium after restoration, December 2003 (courtesy of Teatro La Fenice, venice)

References

1. P. Morachiello, M. Piana, La Fenice un mese dopo. In: La Nuova Venezia, 1996.
2. The quotations are from M.T. Muraro, Gran Teatro la Fenice. Corbo e Fiore, Venezia, 1996.
3. The quotations are from M.T. Muraro, Gran Teatro la Fenice. Corbo e Fiore, Venezia, 1996.
4. A. Baricco, Storia di un teatro che visse due volte. In: La Repubblica 2003; republished in: Teatro la Fenice. Splendidezza di ornamenti e dorature. De Luca editore, Roma, 2003.

Homage to Andrea Pozzo

Exactitude and Extravagance:
Andrea Pozzo's "Viewpoint"

Filippo Camerota

Andrea Pozzo's perspective inventions represent the art of *quadratura* on its high-est level. The illusory power of his fictive architecture was so strong that it force-fully conditioned the perception of space, leading neo-classical critics to censor his paintings as if they were actually buildings. Eloquent on this subject is the criticism of Francesco Milizia, who warned young architects not to follow the example of Brother Pozzo, calling him an "architect in reverse"[1]. On the technical level, in-stead, the supreme skill of the Jesuit painter was undeniable. He could in fact be criticized for one thing only, his decision to impose a compulsory viewpoint on the observer. But it was just this point that formed the cornerstone of his art, and Pozzo, heedless of the critics, never abandoned that crucial requisite.

On both the theoretical and the practical levels, the meticulous exactitude of An-drea Pozzo was impeccable. The didactic structure of his treatise, leading the reader to tackle a problem only after having thoroughly understood the one before it[2], is perfectly matched by his methodical procedure in his worksite, recognizable today by some meaningful signs. On the ceiling of the church of *St. Ignatius* in Rome, for instance, the traces of the construction grid cut into the plaster reveal the utmost degree of precision in transferring the measurements of the sketch to the great di-mensions of the pictorial surface (Fig. 1). On the construction site, as in the treatise, each step is accurately measured, and in this absolute precision lies the secret of the amazing illusionistic effects achieved by Pozzo.

His precision in drawing was undoubtedly a natural talent, and a mere glance at his freehand sketch of the Colosseum in perspective shows that the hand and eye of Andrea Pozzo were guided by a perfect sense of proportion and order (Fig. 2). This natural talent, however, became the operational principle that the painter demanded of his pupils as well, and tried to teach anyone who ventured on the study of per-spective. Priority was categorically assigned to setting the "hand to the compass, and to the rule", rather than indulging in theoretical speculation[3]. Such precision, moreover, found an ideal context in the scientific culture represented on the highest

Filippo Camerota
Museo Galileo, Florence (Italy).

[1] [24], Book III, p. 276.

[2] Pozzo (1693-1700), Part I (1693), p. 12, *Avvisi a i principianti*. Pozzo (1725), *Advise to Beginners*.

[3] Pozzo (1693-1700), Part II (1700), pp. 19-20.

Emmer M. (Ed.): Imagine Math. Between Culture and Mathematics
DOI 10.1007/978-88-470-2427-4_4, © Springer-Verlag Italia 2012

Fig. 1 A. Pozzo, *Gloria di Sant'Ignazio*, Rome, Church of St. Ignatius, 1685: detail of the grid carved in the plaster

level throughout the Christianized world by the mathematicians of the Jesuit order. Years before, French studies in perspective had been embroiled in an intellectual dispute between the Parisian Jesuit Jean Dubreuil and the Lyonese mathematician Girard Desargues. Although Pozzo's theoretical work does not reflect this controversy, the bases for his preference for the single observation point seem to be rooted expressly in that issue.

Fig. 2 A. Pozzo, *View of the Coliseum*, preparatory drawing for the Fig. 44 (*Perspectiva*, II), Rome, Valentino Martinelli Collection

Criticism of the single viewpoint derived from the well-established operational method that held it a good rule to adopt several vanishing points in perspective compositions on ceilings and vaults, in order to ensure the best visibility of the work[4].

This rule, although conflicting with the principles of linear perspective, favoured observation of the painting from the four sides of the room, reducing to a minimum marginal deformations, commonly considered defects to be eliminated. Andrea Pozzo's method reversed this assumption, by imposing a single observation point and bringing marginal deformations to their extreme consequences. In his treatise, this method is disguised by a kind of propaganda typical of the Jesuits ("my advice is, that you chearfully [sic] begin your Work, with a Resolution to draw all the Points thereof to that true Point, the Glory of GOD")[5]. But regardless of its rhetorical form, his motivation appears firmly anchored to a precise objective, that of exaggerating perspectival deformations at the margins of the painting to make the illusion more spectacular from the chosen viewpoint; and thus amazing the observer still further by unveiling the deception.

On this subject Pozzo concurred with his charismatic superior, the Jesuit Father Athanasius Kircher, who used to entertain visitors in his famous museum with erudite scientific and philosophical discussions only after having revealed the trickery of his magic optical games. When Pozzo went to Rome in September 1681, Kircher had been dead for almost a year but his museum was still very much alive. The angel on the ceiling of St. Ignatius holding a concave mirror to reflect the Divine image in the form of a Christogram is presumably a sign of the artist's fascination with Kircher's catoptric machines (Fig. 3). Pozzo could have learned from the curator of the museum, if not from the Jesuit priest's books, that the concave mirror not only lights a fire (symbolic of Ignatius igniting mankind with faith) but also projects images beyond its surface, which appear suspended in mid-air (symbolic of Ignatius reflecting Divine glory). The optical phenomenon described and handed down over the centuries in the writings of Hero of Alexandria, Roger Bacon and, more recently, Giovanni Battista della Porta, had been illustrated by Kircher in some of the most interesting pages of the *Ars magna lucis et umbrae* (*The great art of light and shadows*, 1646)[6]. The Christogram painted by Pozzo appears, in fact, suspended in mid-air, as the double of the image appearing on the surface of the mirror. It is a refined optical effect hard to appreciate from below, but obviously worthy of representation as the figurative expression of an optical theorem. On the symbolic level then, the concave mirror was a true scientific icon, the fundamental instrument of mediation between Divine wisdom and human knowledge, as demonstrated in the

---

[4] On the history of quadraturism, see [25], XI, pp. 99-116; [33], [21], [17].

[5] Pozzo (1693-1700), Part I (1693), p. 13, *Al lettore studioso di Prospettiva*; Pozzo (1725), *To the Lovers of Perspective*.

[6] See [1], [11], [16]. On this subject see also [3].

Fig. 3 A. Pozzo, *Gloria di Sant'Ignazio*, Rome, Church of St. Ignatius, 1685: angel with a concave mirror

Fig. 4 Zacharias Traber, *Nervus opticus*, Wien, 1675, frontispiece

frontispieces to some of the major Jesuit treatises on optics, from Kircher's *Ars magna* to Zacharias Traber's *Nervus opticus* (Fig. 4).

In Rome, Pozzo found the ideal climate for his creative vein. The Monastery of the Minimi at Trinità dei Monti, where he worked, had not only been central to the scientific debate nourished by Galileo's discoveries and Descartes's philosophy, but was also the most advanced centre for studies of optics and its applications. The two great works of anamorphosis painted by the Minim friars Emmanuel Maignan and Jean François Niceron in the cloister of their monastery aroused the same amazement as Kircher's catoptric machines. In these paintings Pozzo could find a direct counterpart to his art. In them, the figures of saints seen from a single, highly off-centre viewpoint slowly dissolved to the point of disappearing as the observer gradually moved toward the centre of the painting. In losing their original form, the paintings took on another, leaving to the observer the pleasure of discovering the secret mechanism of that magical artifice[7].

In walking along the corridor of the Jesuits' *Casa Professa*, Brother Pozzo's first Roman work, the sensation produced by the works of anamorphosis of Trinità dei Monti is felt again. As the visitor proceeds toward the centre of the room, its image is transformed. From the central station point, the corridor appears in all its architectural grace (Fig. 5). Toward the end and toward the entrance, the space is rhythmically marked by a series of architraves supported by corbels resting on projecting pilaster strips, with windows and doors between them. Everywhere, angels and cherubs enliven the scene and a great serliana dramatically frames the altar of *St. Ignatius* in the background. But in proceeding toward the altar, this graceful composure gives way to chaos. The architraves curve into fluidity (Fig. 6); the pilaster strips widen, becoming more oblique; the angels are deformed to the limits of anamorphosis (Fig. 7); and the columns of the serliana slide dangerously down a steep floor (Fig. 8). Faced with the aberration of these forms – too exaggerated not to appear as the most sophisticated virtuosity – the observer, amazed at this supreme display of artifice, retraces his steps to find again, almost unbelievably, the lost grace and composure.

Fig. 5 A. Pozzo, Corridor of the rooms of St. Ignatius, Rome, Casa Professa del Gesù, 1682, view from the center of the corridor

---

[7] On this subject see [4], pp. 51-75.

Fig. 6 A. Pozzo, Corridor of the rooms of St. Ignatius, Rome, Casa Professa del Gesù, 1682, deformation of the architraves towards the end of the corridor

Fig. 7 A. Pozzo, Corridor of the rooms of St. Ignatius, Rome, Casa Professa del Gesù, 1682, perspective deformation of the figures towards the end of the corridor

Fig. 8 A. Pozzo, Corridor of the rooms of St. Ignatius, Rome, Casa Professa del Gesù, 1682, frontal view of the oblique wall at the end of the corridor

The corridor of the *Casa Professa* is an emblematic case study of the effects produced by a single viewpoint. Before such a long, narrow and relatively low room, any quadraturist would have applied the rule of multiple viewpoints, unless he was trying – as Pozzo was – to exploit perspective deformation to heighten the impact of the visual experience. "And although it is possible to divide walls, or very long, low ceilings, into several parts in the work", writes Pozzo, "and to assign its own observation point to each of them, it nonetheless appears that a much more ingenious effect can be attained in such cases using a single viewpoint, as I did in a corridor of the Church of *Jesus* in Rome"[8]. The striking distortion seen when the image was observed from any other vantage point was, Pozzo declared, "not a defect but praiseworthy art". Deformation of this kind was not restricted to the art of painting alone. The Baroque century had seen the triumph of fluid forms in architecture, and Rome had been its great experimental theatre, on both the practical and the theoretical levels. Borromini's buildings quivered with real and latent aberrations: curved and undulating architraves, concave and convex walls, winding spirals, leaning arches; oblique forms whose theoretical counterparts could be found in the bizarre mathematical hypothesis of Juan Caramuel de Lobkowitz known as *architettura obliqua*[9].

In the garden of *Palazzo Spada* is a work that might have struck Pozzo's fancy. Here Borromini designed a perspective gallery, ingeniously built by the Augustinian

---

[8] Pozzo (1693-1700), Part I, ed. 1717, Fig. 101, pp. 216-217.

[9] See [9].

Fig. 9 Oblique deformation of the architectural orders in the perspective gallery of Palazzo Spada, Rome

monk Giovanni Maria da Bitonto, which gave the illusion of connecting the secret garden to a larger garden, one entirely non-existent (Fig. 9). A *material* perspective, of the type Pozzo had designed for numerous theatrical scenes and fictive altars, it was not made of wood and canvas but of masonry; and this made it something very different from a mere ephemeral stage-set. Its beauty lay, and still lies today, not in the image that appears from the preferential viewpoint, but in the experience of walking through it to discover how the space and the elements in it are gradually transformed[10]. In this case too, the deception is only enhanced by being revealed.

Confronted with this work, Pozzo's thoughts may have gone back to his first years in Milan, when, in studying masterpieces of perspective "in the academies, the galleries and the churches"[11], he must have come across the first material perspective ever built in masonry: the illusory choir of *Santa Maria presso San Satiro*, the famous artifice employed by Bramante to compensate for the lack of sufficient space to build a real architectural structure (Fig. 10). The great master of classical Renaissance architecture had used perspective in an innovative way, indicating how visual deception could be used to solve architectural problems. It may have been this first experience that induced Andrea Pozzo to go beyond temporary stage sets and ephemeral decorations, seeking to attain an optical dimension in architecture. In

[10] See [26]; see also [18]; [8], pp. 293-295.
[11] Baldinucci (1975), p. 316.

Fig. 10 Oblique deformation of the architectural orders in the fake choir of Santa Maria presso San Satiro, Milan

his treatise, the problems of perspective are discussed beginning with a similar case, the construction of a fictive choir in a church, establishing the principle that "the Perspective of Structures here treated of, can have no Grace or Proportion, without the Help of Architecture" (Fig. 11)[12].

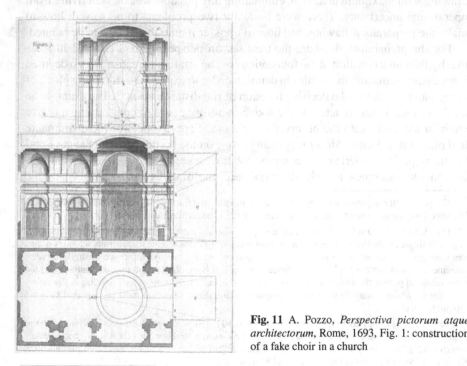

Fig. 11 A. Pozzo, *Perspectiva pictorum atque architectorum*, Rome, 1693, Fig. 1: construction of a fake choir in a church

---

[12] Pozzo (1693-1700), Part I (1693), p. 12, *Avvisi a i principianti*. Pozzo (1725), *Advise to Beginners*.

According to this principle, already expressed by Sebastiano Serlio, who saw it applied on the highest level by Bramante, the painter must necessarily be an architect as well[13]. Art's relationship to architecture was strengthened by the role of perspective which was used mainly to design architectural settings in paintings, stage-sets and temporary decorations for festivals and processions. In addition to the rules of geometric drawing, then, the artist was obliged to learn the rules of architectural drawing: first of all orthogonal projections, but also the proportions of the orders. Pozzo was certainly familiar with Jean Dubreuil's influential treatise *La Perspective pratique* (*Practical Perspective*, 1642), where perspective was defined as a "science [that] can boast of being the life and soul of painting" and that reached its apex of beauty in representing "rich, sumptuous buildings, constructed according to the orders of columns, whose beauty depends on the proportions and dimensions that must be respected to avoid wounding the eye"[14]. Hence proportions and dimensions are the indispensable underpinnings for conferring beauty on a perspective composition. The fact that Pozzo's frescoes aroused debate of an architectural nature shows how far beyond the limitations of mere pictorial decoration his art had gone[15].

The importance assigned by Pozzo to the two fundamental components of perspective drawing, *proportions* and *dimensions*, is evident from his rigorous mode of guiding the reader of his treatise to the perfect execution of a work. He is not interested in explaining the geometrical optical principles, presuming that his readers are already familiar with the existing texts, but concentrates instead on constructing a drawing with maximum precision, eliminating any possible weakness deriving from operational uncertainty. There were basically two problems to be solved: how to make the preparatory drawing, and how to transfer it onto the surface to be painted.

For the preparatory drawing, the treatises on perspective had established rules that by then now constituted the foundations of the art. In his treatise, Pozzo deemed it necessary to discuss these rules in detail in order to explain the stumbling block of many painters, that is, the decisive function of the distance point[16]. The example he used to illustrate this function is the above-mentioned case of constructing a fictive choir in a church, with the observer's eye located approximately where Bramante had placed it in *Santa Maria presso San Satiro*, on the border between the nave and the transept. The generic cases common to all treatises on perspective are followed by concrete examples, namely, the works accomplished by Pozzo himself "by the

---

[13] [32], p. 25v: "Il perspetivo non fara cosa alcuna senza l'Architetura, ne l'Architetto senza perspettiva" [Perspectivist can do nothing without architecture, nor can the architect without perspective].

[14] [13], Part I (1642), *Preface*: "Cette science se peut vanter d'estre l'ame et la vie de la Peinture [...] le plus belle pieces de Perspective, se font de Bastimens riches et somptueux, construits selon les ordres des Colomnes, la beauté desquels dépend des proportions et des mesures qui doivent estre observées, autrement elles blesseront l'oeil" [This science can boast of being the life and soul of Painting [...] the most beautiful perspectives are made with rich, sumptuous buildings, constructed according to the orders of columns, whose beauty depends on the proportions and dimensions that must be respected to avoid wound the eye].

[15] Pozzo (1693-1700), Part I (1693), Fig. 91, pp. 196-197: "Si meravigliarono alcuni architetti, che io appoggiassi le colonne davanti sopra mensole, ciò che essi non farebbono in una fabrica vera, e reale". Pozzo (1725), The Ninety-first Figure: "Some Architects dislik'd my setting the advanc'd Columns upon Corbels, as being a thing not practis'd in solid Structures".

[16] Pozzo (1693-1700), Part I (1693), Fig. 1, p. 16. Pozzo (1725), The First Figure.

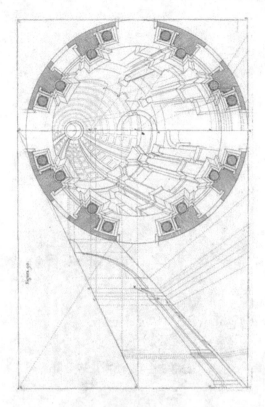

Fig. 12 A. Pozzo, *Perspectiva pictorum atque architectorum*, Rome, 1693, Fig. 90: perspective construction of the dome in the Church of St. Ignatius, using the ordinary rule (distance point)

Rule I make use of at present", deemed "more easy and general than the common way"[17]. The *common* rule consisted of construction with the distance point, in which the volumes of bodies were obtained by representing the two orthogonal projections, plan and elevation, in perspective (Fig. 12). The *more easy* rule, explained in the second part of the treatise, consisted of intersecting the visual pyramid in plan and elevation, thus relieving the painter of the projective complications involved in the former method (Fig. 13).

The problem of transferring the preparatory drawing onto the surface to be painted was even more complex, since the drawing had to be enlarged and adapted to the shape of the picture plane. In this case the procedure employed was *quadrettatura*, or gridding, the most precise of the many methods devised over the years by painters and mathematicians. In fictive altars, in theatrical scenes and in temporary decorative sets for the *Quarant'ore*, or Forty Hours devotion, where the perspective was painted on a series of vertical frames placed one after another, the so-called backdrops, the problem consisted basically of how to draw with precision the proportional variation of the grid on the various planes of the painting so that, from

[17] Pozzo (1693-1700), Part I (1693), p. 13, *Al lettore studioso di Prospettiva*. Pozzo (1725), *To the Lovers of Perspective*.

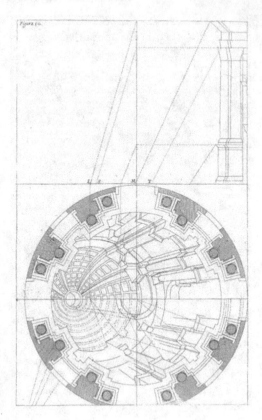

Fig. 13 A. Pozzo, *Perspectiva pictorum atque architectorum*, Rome, 1700, Fig. 52: perspective construction of the dome in the Church of St. Ignatius, using the easiest rule (intersection of the visual pyramid in plan and elevation)

the preferential viewpoint, all of the lines appeared to extend in perfect continuity ("You must be very careful that all the Squares of the Net-work be exactly divided, and at right Angles")[18].

Being at different distances from the eye, in fact, the squares of the grid drawn on the various planes of the painting differed in size (Fig. 14).

Even more difficult was the "method for marking grids on vaults", where the concave surface, frequently interrupted by ribs or groins, called for controlled deformation of the grid. Theoretically, it was sufficient to construct a network of taut cords stretched over the impost plane of the vault and to project their shadows by means of a lantern.

Then the painter would only have to trace the shadow with his brush, as described in some recent widely-read treatises, such as those of Abraham Bosse and Gregoire Huret (Fig. 15). But this was only theoretical. "I say, *if you imagine* a Lamp thus fix'd", wrote Pozzo in explaining the concept, "because either the Scaffold to the Vault, or the great Distance of the Vault from the Net-work, or the greater of both

---

[18] Pozzo (1693-1700), Part I (1693), Fig. 62, p. 138: *Del graticolare i telari che rappresentano fabriche di rilievo*. Pozzo (1725), The Sixty-second Figure, *Of making the Net-work on Frames, for representing the Architecture as solid*.

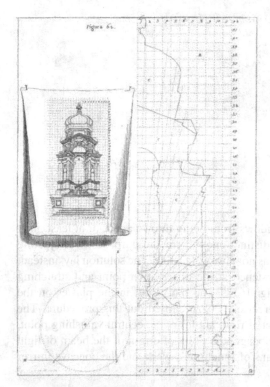

Fig. 14 A. Pozzo, *Perspectiva pictorum atque architectorum*, Rome, 1693, Figs. 61-62: constructive grid for a fake altar

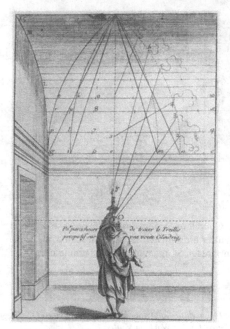

Fig. 15 A. Bosse, *Moyen universale pour pratiquer la perspective sur les tableaux ou surfaces irregulieres*, Paris, 1653, pl. 15: tracing perspective lines on the surface of a vault

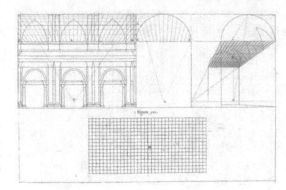

Fig. 16 A. Pozzo, *Perspectiva pictorum atque architectorum*, Rome, 1693, Fig. 100: tracing the grid on the surface of the vault in the Church of St. Ignatius

from the Light, may prevent the Shadows from being thrown at all, or at least, may render them so faint, as not to be distinct enough for the purpose"[19]. Projecting shadows, although a brilliant idea, was not always feasible. The solution lay, instead, in using a long rope "as a Ray", fastened at the observation point and stretching up to the surface of the ceiling (Fig. 16)[20]. The marks cut in the plaster on the ceiling of St. Ignatius reveal how precisely Pozzo carried out this procedure. The two middle lines in the grid intersect at right angles at the central vanishing point, emblematically placed on Christ's ribcage, the point of origin of the beam of light conveying the Divine image by means of Ignatuius's work and the concave mirror (Fig. 17).

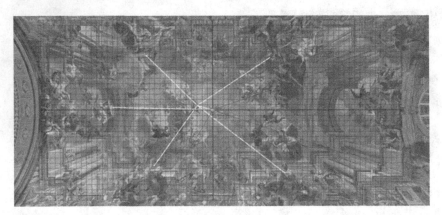

Fig. 17 A. Pozzo, *Gloria di Sant'Ignazio*, Rome, Church of St. Ignatius, 1685: reconstruction of the grid traced by Pozzo (the lines carved in the plaster are still visible). The white lines show the light rays which, from Christs ribs (vanishing point), reach St. Ignatius and the burning mirror

---

[19] Pozzo (1693-1700), Part I (1693), Fig. 100, p. 216: *Modo di far la graticola nelle volte*. Pozzo (1725), The Hundredth Figure, *The Method of drawing the Net or Lattice Work on Vaults*.

[20] Ibidem.

Once the grid had been traced, the artist had to control with maximum precision the lines converging at the central vanishing point, which had to appear straight although they were of course curved on the ceiling. For this purpose, two cords fastened at the centre of the ceiling, at the vanishing point, were used: one "to guide the rule straight" in tracing the lines, the other left "hanging like a pendulum" to sight and correct any deviations of the rule[21]. In geometric terms, the two cords were two straight lines belonging to the same vertical plane passing through the observer's eye, whose intersection with the ceiling generated a curved line.

Although applied mainly to the problems of artists, and thus consisting of practical procedures, perspective was by tradition a mathematical science. Christoph Clavius had included it in the curriculum of Jesuit schools, and its applications had been studied by some of the Company's illustrious mathematicians, such as Christoph Scheiner, inventor of the pantograph, Mario Bettini and Christoph Grienberger, inventors of new perspective instruments inspired by the pantograph, Athanasius Kircher and Gaspard Schott, visionary creators of magic optical tricks, and Jean Dubreuil, one of the leading figures in the heated debate that accompanied the development of projective geometry in France[22]. Pozzo was undoubtedly familiar with that debate, since the issue had a direct bearing on the teaching of perspective in the Jesuit colleges, downplaying the importance of the distance point, which for Pozzo, as mentioned, was the thing *most necessary* to the perfect execution of a painting.

The bitter academic quarrel had been unwittingly triggered by Dubreuil himself. In his *Perspective pratique*, a text widely disseminated and not only in Jesuit schools, he had presented with some variants the perspective method developed a few years earlier by the architect and mathematician Girard Desargues (1642, pp. 117-119). Desargues had reacted vehemently, affixing handbills in the streets of Paris publicly accusing the Jesuit father of plagiarism and proclaiming him guilty of "unbelievable errors" and "enormous mistakes and falsehoods"[23].

Between 1636 and 1640, Desargues had published three important treatises – on perspective (1636); on conic sections (1639); and on stereotomy (1640) – proposing a radical renovation of the methods used for representing geometric figures. For Desargues, perspective and geometric drawing (isometric drawing and orthogonal projections) were "two species of the same gender"[24] that depended on a general method aimed at measuring the geometric position of all points through their spatial coordinates.

According to Desargues, Dubreuil had failed to grasp the fine points of his *maniére universelle*, and had trivialized it in the pages of his *Perspective pratique*. To these accusations, Dubreuil had replied with a scornful pamphlet, *Advis charitables (Charitable Advice)*, and a new text published immediately, *Diverses methodes universelles et nouvelles (Various new and universal methods)*, attributing the au-

---

[21] Ibidem.

[22] See [30]; [5], Apiarium V, Progymnasma II, pp. 35-56; [22], Book II, Part II; Book X, Part II; [31], Book III; [14].

[23] On this subject, see [29]; [4], pp. 87-94.

[24] See [12].

thorship of Desargues's *maniére* to the mathematician Jacques Alleaume. In the meantime, the controversy had produced several texts on perspective by the most eminent members of the Académie royale in Paris, fiercely attacking one another; on one side, Abraham Bosse, untiring defender of Desargues; on the other, Jacques Curabelle, Jacques Le Bicheur and Grégoire Huret, who spared their opponents no insult nor personal offence[25].

The *Maniére universelle de M. Desargues* (*The Universal Method of Mr Desargues*) by Abraham Bosse was published in two successive volumes, the second of them devoted to a *Moyen universel de pratiquer la perspective sur les tableaux ou surfaces irrégulières* (*Universal method for drawing in perspective on paintings or irregular surfaces*, 1653), that is, a method for painting on vaults and domes. The treatise dealt with questions of projection that were exemplified by the perfect practical procedure for transferring a drawing to the surface of a ceiling. To control the deformation introduced by the shape of the pictorial support, Bosse suggested projecting the drawing "with cords, or using candles"[26]. The best method was the one adopted by Pozzo as well, that is, constructing a horizontal network at the base of the ceiling, with cords stretched from one side to the other of the impost plane, and projecting its shadow by means of a candle placed at the observation point (Fig. 18). This procedure was illustrated by Gregoire Huret, who in 1663 had replaced Bosse

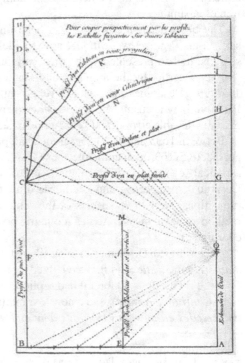

Fig. 18 A. Bosse, *Moyen universale pour pratiquer la perspective sur les tableaux ou surfaces irregulieres*, Paris 1653, pls. 6-7: deformation of the grid on different surfaces

[25] See [6], [10], [23], [20].

[26] [7], pl. 15.

as teacher of perspective at the Académie royale and had made his debut in the literature about art with a text in which he claimed to be able to draw any building in perspective, without using either the plan or the distance point[27]. The object of simplifying the technical procedure while at the same time guaranteeing precision in projection was an attempt to go beyond the method of Desargues who, withdrawing from the debate, had challenged mathematicians and artists to write a text better than his.

Apart from personal malice and academic rivalry, the debate centred on the age-old question of the relationships between the principles of geometry and the practice of art. Desargues and Bosse advocated total geometric control in architectural drawing as well as painting, but to their opponents, the use of proportional scales seemed a futile complication. Dubreuil also deemed the use of geometric *rules fundamental*, since without them the painter could hope "only to paint for the ignorant", but in artistic practice their application was functional and did not call for the rigour of a mathematical demonstration[28]. The French academics were less willing to compromise. Pointing to the recent Paris edition of Leonardo's *Trattato della pittura* (*Treatise on painting*) as authority, they claimed priority for the judgement of the eye.

Pozzo's position in this debate was close to that of Dubreuil's: lose no time "in mere speculation" but take up "the compass and rule", in order to apply the geometric rules with the utmost precision. Only with these instruments could "the most subtle of all our outward Senses [the Eye]" be deceived[29], in architecture as well as figures: "and is very necessary to be known of all, who in Painting would give a due Place and Proportion to their Figures"[30]. Without any theoretical pretensions, entrusting to drawings the role of instructor ("if you meet with any thing which at first seems difficult in the Description, a diligent Inspection of the Figure may relieve you")[31], Pozzo managed to blend operational practice and scientific exactitude with surprising results. In his treatise, mathematical demonstrations are suppressed in favour of a perfect understanding of the method of operating with the rule and compasses, whose foundations are solidly based, however, "in the *Art of Geometry* and of *Architecture*, which I assume are already known to those who undertake this study"[32].

With a rigour that matches his technical skill, Pozzo sets forth the two basic rules of practical perspective, *distance point* and *intersection*, explaining how to avoid the pitfalls of geometric drawing on highly foreshortened planes. With the same inflexible rigour, Pozzo rejects any mitigation of perspectival deformation, such as the use of several vanishing points preferred by the great quadraturists of Emilia and Veneto, firmly imposing his faith in the single vanishing point that will provide

---

[27] See [19].

[28] [13], Part I (1642), *Preface*.

[29] Pozzo (1693-1700), Part I (1693), *Al lettore studioso di Prospettiva*. Pozzo (1725), *To the Lovers of Perspective*.

[30] Pozzo (1693-1700), Part II (1700), *Al lettore*. Pozzo (1725), *To the Lovers of Perspective*.

[31] Pozzo (1693-1700), Part I (1693), p. 12, *Avvisi a i principianti*. Pozzo (1725), *Advice to Beginners*.

[32] Pozzo (1693-1700), Part II (1700), *Al lettore*.

maximum illusion, and assigning an important added value to marginal deformations. To this crucial subject, Pozzo devoted the concluding appendix of his treatise, a *Risposta sull'obiezione fatta circa il punto di vista nelle prospettive* (An Answer to the objection made about the Point of Sight in Perspective), where he sets forth three simple and categorical points: *first*, "the greatest Masters" have always used a single viewpoint; *second*, "since perspective is but a Counterfeiting of the Truth" the painter is not obliged to show it correctly from all points of view, but from only one; *third*, if a work is made to be seen from several viewpoints, in none of them will the illusion be truly convincing. The glorious fresco over the nave in St. Ignatius is cited as proof of these statements, and the marble disk in that church that unmistakeably marks the position of the observation point has the indisputable truth of a geometric theorem, one whose proof is repeated anew each time to the astonished eyes of the observer.

References

1. D.C Lindberg, Roger Bacon's philosophy of nature. A critical edition, with English translation, introduction and notes, of De multiplicatione specierum and De speculis comburentibus. Clarendon Press, Oxford, 1983.
2. F.S. Baldinucci, Vite di artisti dei secoli XVII-XVIII: prima edizione integrale del Codice Palatino 565, (ed.) Matteoli A. Roma, 1975, pp. 314-337.
3. J. Baltrušaitis, Lo specchio. Rivelazioni, inganni e science fiction. Adelphi, Milano, 2007.
4. J. Baltrušaitis, Anamorfosi, o, Thaumaturgus opticus. Adelphi, Milano, 1990.
5. M. Bettini, Apiaria universae philosophiae mathematicae. I.B. Ferroni, Bologna, 1645.
6. A. Bosse, Manière universelle de G. Desargues pour pratiquer la perspective. Pierre Des-Hayes, Paris, 1643.
7. A. Bosse, Moyen universel de pratiquer la perspective sur les tableaux ou surfaces irrégulières. Paris, 1653.
8. F. Camerota, La prospettiva del Rinascimento. Arte architettura scienza. Electa, Milano, 2006.
9. J. Caramuel de Lobkowitz, Architectura civil recta y obliqua, considerada y dibuxada en el Templo de Jerusalem. Camillo Corrado, Vigevano, 1678.
10. J. Curabelle, Examen des oeuvres du Sieur Désargues. M.&I. Henault, Paris, 1644.
11. G.B. Della Porta, Magiae Naturalis libri XX. Matthiam Cancer, Napoli, 1589.
12. G. Desargues, Exemple d'une manière universelle de S.G.D.L. touchant la pratique de la perspective. Bidault, Paris, 1636.
13. J. Dubreuil, La perspective pratique necessaire a tous peintres, graveurs, sculpteurs, architectes, orfeures, brodeurs, tapissiers, & autres se servans du dessein. Melchior Tavernier, Paris, 1642-1647-1649.
14. J. Dubreuil, Advis charitables sur les diverses oeuvres, et feuilles volantes du sr. Girard Desargues Lyonois. Melchior Tavernier, Paris, 1642.
15. J. Dubreuil, Diverses methodes universelles et nouvelles en tout ou en partie pour faire des perspectives. Melchior Tavernier, Paris, 1642.
16. Hero of Alexandria, Mechanica et catoptrica. In: Heronis Alexandrini Opera quae supersunt omnia. Teubner, Stuttgart,1976.
17. E. Filippi, L'arte della prospettiva. L'opera e l'insegnamento di Andrea Pozzo e Ferdinando Galli Bibiena in Piemonte. Olschki, Firenze, 2002.
18. M. Heimbürger Ravalli, Architettura, scultura e arti minori nel Barocco italiano. Ricerche nell'Archivio Spada. Palombi, Firenze, 1977.
19. G. Huret, Exposé d'un régle precise por decrier le profil élevé du fut des colonnes. Paris, 1665.

20. G. Huret, Optique de Portraiture et Peinture. Huret, Paris, 1670.
21. M. Kemp, The science of art. Optical themes in Western art from Brunelleschi to Seurat. Yale University Press, New Haven, 1990.
22. A. Kircher, Ars Magna Lucis et Umbrae. Grignani, Roma, 1646.
23. J. Le Bicheur, Traicté de Perspective. Jollain, Paris, 1660.
24. F. Milizia, Memorie degli architetti antichi e moderni. Stamperia Reale, Parma, 1781.
25. F. Negri Arnoldi, Prospettici e quadraturisti. In: Enciclopedia Universale dell'Arte, vol. XI, pp. 99-116. De Agostini, Novara, 1963.
26. L. Neppi, Palazzo Spada. Edizioni d'Italia, Roma, 1975.
27. A. Pozzo, Perspectiva pictorum atque architectorum. Komarek, Roma, 1693-1700.
28. A. Pozzo, Rules and examples of perspective proper for painters and architects. London, 1725.
29. N.G. Poudra, Oeuvres de Desargues reunies et analiste. Leiber, Paris, 1864.
30. C. Scheiner, Pantographice, seu ars delineandi res quaslibet per parallelogrammum lineare seu cavum, mechanicum, mobile. Grignani, Roma, 1631.
31. G. Schott, Magia universalis naturae et artis, sive, Recondita naturalium & artificialium rerum scientia. Schönwetter, Würzburg, 1657.
32. S. Serlio, Il Primo libró d'Architettura [di Geometria]. Il Secondo Libro di Perspettiva. Barbé, Paris, 1545.
33. N. Spinosa, Spazio infinito e decorazione barocca. In: Storia dell'arte italiana, 6, Einaudi, Torino, 1981, pp. 275-343.

The Apse Scenes in the Prospective Inventions of Andrea Pozzo

Silvia Carandini

It was the great multi-talented artists such as Bernini, and Giovan Battista Aleotti, Bernardo Buontalenti and Giacomo Torelli before him, using a toolkit that included a vast array of figurative languages as well as the mechanical and optical sciences, who wove the threads of an extraordinary network of evocations and borrowed devices that were distilled from the immense cauldron of literature, arts and theatre of the Baroque period. For all these artists, works in architecture and in theatre constituted crucial experiences in the profession, the laboratories for comparing artistic forms of expression and scientific practices. At the end of the seventeenth century, a significant role was played by Andrea Pozzo (Trento 1642 – Vienna 1709). A simple Jesuit brother, he occupied an important place, gathering to himself the entire legacy of his predecessors, although employing it almost exclusively in a religious context ([2]; [7]; [9]). His magnificently illustrated treatise, *Perspectiva pictorum et architectorum*, published in Rome in two volumes by the typographer Komarek between 1693 and 1700, is the most evocative and efficacious document possible of praxis and mastery; here Pozzo presents the sum of his experiences as a painter, architect, decorator and scene designer, all united in the service of that *perspectiva artificialis* which over the course of the 1600s had benefitted from the contributions of so many scientists and artists. Andrea Pozzo had also contributed to the perfection of this field of practical experimentation; he, like the other great artists of the 1600s, was commissioned to create decorations and backdrops for festive occasions, and relied on temporarily-erected mechanical contrivances to create sacred theatrical spaces for religious ceremonies. His talents had also been refined through designs for the ornamentation of churches, such as tabernacles and altars, both on paper and actually built, of solid and precious materials, or more economically and with astonishing skill, in faux materials and paint. He also had occasion to test his skills in theatrical designs for the extremely active stages in the Jesuit colleges, in particular those of Milan and Rome.

All of these aspects of Pozzo's work have been the objects of recent studies occasioned by the three-hundredth anniversary of his death, celebrated in 2009-2011. Particularly noteworthy are the studies of Father Heinrich Pfeiffer, Francesco Camerota, Elysabeth Kieven, Richard Bösel, Lidia Salviucci Insolera, Giuseppe

Silvia Carandini
Department of History of Art and Performing Arts, Sapienza University of Rome (Italy).

Emmer M. (Ed.): Imagine Math. Between Culture and Mathematics
DOI 10.1007/978-88-470-2427-4_5, © Springer-Verlag Italia 2012

Dardanello and Marinella Pigozzi ([3], [4], [11]), and of Pascal Dubourg Glatigny in the recently published anastatic reprint of his treatise ([12]).

A frontispiece of the second volume of the *Perspectiva pictorum et architectorum* may perhaps furnish the key to understanding this humble yet ardent Jesuit brother (Fig. 1).The engraving shows a benevolent Minerva offering refreshment to an artist who kneels down, and drawing water from a well (pozzo) enshrined in a small apse framed by a double arch, between columns and pilasters. We might call this ingenious work a dual self-portrait, in which he depicts himself (Pozzo) as modestly drinking from the science of the ancients, and at the same time, as an inexhaustible well (pozzo) of knowledge, which his treatise freely imparts.

It is with this small apse framing the 'well of science', as though it were an altar, that I would like to begin this brief delineation of an itinerary through the works of Andrea Pozzo, which centres on the multiple variations of a theme held dear by the artist: the creation of an illusionistic space, modelled on the apse of the church, which, situated behind a proscenium arch or a triumphal arch, constitutes the fulcrum of a vision of a fictitious nature, like the backdrop of a stage set.

Fig. 1 A. Pozzo, frontispiece, *Perspectiva Pictorum et Architectorum*, vol. II, Rome, 1700

1 The colonnade and arch that frame the vision

As part of the celebrations of the three-hundredth anniversary of Pozzo's death, the restoration of the Chiesa della Missione in Mondovì, dedicated to St Francis Xavier and decorated by the artist, then a young man, between 1675 and 1677, before he was called to Rome, was carried to completion. Using decorations in stucco, faux finishes and frescoes, the artist drapes a brilliantly coloured mantle over Boeri's architecture, transforming the incompletely resolved architectonic composition into a fascinating spectacle, an extremely effective dramatization of the person and gestures of St Francis Xavier, who, together with St Ignatius, founded the Jesuit order. A recently published book documents the magnificence of Pozzo's creation ([11]). Here, restored to all their original splendour, are the rows of columns in rose-coloured faux French breccia marble, standing out against the white of pedestals, pilasters, capitals, and cornices that accompany the spectator's view along the nave, all the way to the altar (Fig. 2). In the centre of all this rises the astounding altar apparatus, the only one of the many temporarily-erected machines contrived by Andrea Pozzo still in existence, now admirably restored. The entire composition is crowned by the illusionistic depictions of the apse and its vault, where the artist's perspective genius was made evident in all its glory for the first time, giving us imaginary views into elevated platforms, doubling the structure of the nave, opening up into non-existent cupolas, virtual heavens, flights of angels, musical concerts, and the mystical vision of the saint himself ascending into paradise.

2 Imaginary apses and painted altars

Another of Andrea Pozzo's remarkable examples of illusionism is the daringly foreshortened painting at the end of the well known corridor leading to the quarters of

Fig. 2 Mondovì, Chiesa della Missione, view of the nave

St Ignatius in the *Chiesa del Gesù* in Rome, in which, following a series of modillions, architraves and arches, there appears an apse with two angels playing musical instruments who introduce the scene. At the centre of the altar is an aedicule, as though in relief, with columns and broken pediments, and at the focal point of the scene is the altarpiece of St Ignatius. The view that is seen when the spectator stands in the centre of the corridor deforms as the spectator moves closer to the wall, precipitating into a chaos of profoundly altered forms, where reality is decomposed, as in the enigmatic experience of anamorphosis ([5]).

While in this case astonishment comes into play, the mystery of a reality that is revealed only through enigmas, and through an evident disillusionment that increases the force of the illusion ([5]), there were numerous altars that Pozzi created in materials both poor and precious, or, as in the corridor of St Ignatius, painted with an illusionistic technique in churches and then depicted in his treatise, placing his optical prowess at the service of both the glory and the economy of the Jesuit order. Through the use of faux imitations of much more costly materials, decorative elements, statues, paintings on canvas and, stuccos, Pozzo, thanks to the use of chiaroscuro and perspective, provided at a small cost the required depth of altar tables, executed as imaginary projections, as in the Chiesa del Gesù at Frascati, where he creates the illusion of the concave back wall of a large choir in an apse crowned with a cupola on the concave wall of the tribune, while in the middle of the illusory space the convex body of an altar in the form of a tempietto with its little cupola on top *materializes* (Fig. 3); in his treatise he explains with evident pleasure the process used to create the virtuoso and rhetorically *shrewd* double illusion.

Pozzo also uses the play of perspective and the two-dimensional illusion of real space as trial runs for design projects, as in the immense painted canvas depicting a possible solution for the decoration of the tribune in the church of Sant'Ignazio, documented in an etching by Nicolas Dorigny and Antonio Colli ([4], 120). In this case the aim was to make it possible to evaluate the effect of the planned architecture *in loco*, and thus the painting on canvas was hung like an immense stage curtain or backdrop.

3 Temporary and permanent devices

Let us now go back to the nave of the Chiesa della Missione in Mondovì, and in particular to that apse where the white and red sequence of architectural elements terminate (Fig. 4). The altar in the centre was conceived by Pozzo, perhaps provisionally, as a temporarily-erected apparatus in faux materials, shaped like a quadrangular tempietto, with cross-shaped columns, a rich pediment, the space in the middle prepared for a genuine apparition of the image of St Francis Xavier. Engraved and painted on a sheet of metal, the figure was set into motion by a hoist located above, by means of a winch. The effect created was *wondrous*: a vision of the saint framed against the black background, appearing to rise up and levitate in space, in re-enactment of one of the miracles that had contributed to his canonisa-

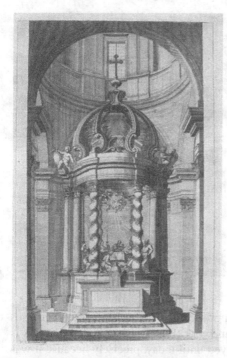

Fig. 3 A. Pozzo, painted altar in Frascati, *Perspectiva Pictorum et Architectorum*, vol. II, fig. 69

Fig. 4 Mondovì, Chiesa della Missione, altar apparatus

Fig. 5 Mondovì, Chiesa della Missione, Chiesa della Missione, altar apparatus viewed from the rear

tion. The structure of the apparatus is comprised of two simple, painted canvases (or backdrops) supported on a wooden framework (Fig. 5). I mentioned that this is the only apparatus by Pozzo that has survived until today, precisely because it was intended for temporary use on the altar. It is thus that much more interesting to be able to look at it in relationship to other apparatuses conceived by Pozzo and documented in his treatise, such as that shown in fig. 62 in the first volume, where he explains the *Modo d'alzar le machine che son composte di più ordini di telari* (Way to lift the machines that are composed of several orders of canvas) (Fig. 6). More than once Andrea Pozzo boasted of the simplicity and economy that characterised his temporary apparatuses, which, while apparently lavish, he created for his order at a minimal cost, with particularly effective illusionistic effects.

4 The scene created for the forty-hours devotion

The pages of the *Perspectiva* offer numerous examples of inventions described by the artist as being suitable for the celebration of the *Quarantore*, or forty-hours devotion (and thus intended to be mounted and demounted), or which lend themselves to a dual purpose, such as the *disegno per l'altar maggiore di qualche Chiesa* (design for the high altar of some church) which can also serve as an apparatus for the forty-hours devotion by 'arranging in the middle space several angels on clouds' (Fig. 7). This particular device is very similar to that built for the high altar of the church in Mondovì, and thus the play of true and false, permanent and temporary, is multiplied.

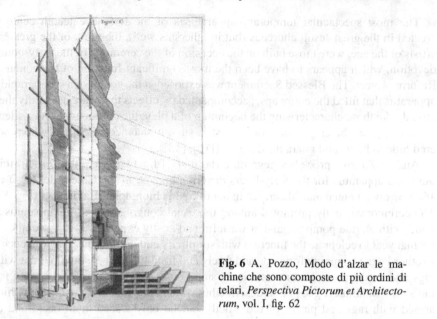

Fig. 6 A. Pozzo, Modo d'alzar le machine che sono composte di più ordini di telari, *Perspectiva Pictorum et Architectorum*, vol. I, fig. 62

Fig. 7 A. Pozzo, Fabrica quadrata, *Perspectiva Pictorum et Architectorum*, vol. I, fig. 64

The most spectacular temporary apparatuses of the late seventeenth century erected in the great Jesuit churches (but in others as well), the work of the greatest artists of the age, were those built on the occasion of the ceremony of the forty-hours devotion, which appears to have been the most significant function of the counter-Reform project. The Blessed Sacrament was exposed at the centre of a scenographic apparatus that filled the entire apse, accompanied by effects that were decidedly theatrical, like those characterising the beginning of a play: the curtain rose and a silent tableau appears, blazingly luminous and still, silently manned by experts who powered hidden lights and guaranteed safety ([10], [13]).

Andrea Pozzo's procedure regarding the first of his two greatest efforts in this field, the apparatus for the forty-hours devotion erected in the Chiesa del Gesù in 1685, is documented and illustrated in detail by his biographer Baldinucci (Fig. 8). The ceremony usually promoted among the usual congregation using apparatuses built 'with all due pomp ... and at wasteful and costly expense ... it is seemly ... for that year to celebrate the function with simplicity and without any magnificence', Brother Pozzo 'stepping forward, said freely and frankly, that he would build it with rags and used canvases: at an extremely small cost, and with increased beauty for the altar'. Baldinucci goes on to describe how others laughed at Pozzo, 'seeing him armed with rags and pieces of old wood', but he only teased them back, and at the end of the work displayed the canvasses stretched out on the ground, the image painted on them incomprehensible without the perspective arrangement it had been designed for. Having created the greatest perplexity in the bystanders, the day after he 'closed himself away with the canvasses where the device was to be installed, he

Fig. 8 A. Pozzo, Teatro delle Nozze di Cana Galilea fatto nella chiesa del Giesu di Roma l'anno 1685, *Perspectiva Pictorum et Architectorum*, vol. I, fig. 62

mounted it all together without showing it to anyone, in order to then uncover it in public at the exposition of the Blessed Sacrament, and so it happened that a most beautiful theatre scene was created before one's eyes in an instant'([1], 322-323).

In the treatise he describes an apparatus called *Teatro delle nozze di Galilea* (Theatre setting for the 'Marriage in Cana'), which he defines as a *nobile architettura*, of a more complex structure than the devices illustrated in the preceding figures. He speaks of six orders of canvasses, 'without counting those which in the middle of the large arch appear to be clouds full of angels in adoration of the Most Blessed Sacrament: and those clouds I have not drawn here, in order not to cover the other parts of the cloths that are further inside' ([12], I: fig. 67-71) .

The other images in the treatise that refer to this particularly scenographic kind of apparatus are very well-known. There was, for example, in 1695, again at the Gesù, the scene entitled *Sitientes venite ad aquas* (Fig. 9). The space of the apse painted just a few years earlier by Giovan Battista Gaulli (called 'il Baciccia') was encapsulated within the apparatus: architecture, painting, and temporary device, here too the functions were interchanged, and real and illusory spaces were combined in an ingenious invention, without ever leaving the confines of the apse, the *scena absidale*.

Fig. 9 A. Pozzo, Teatro tutto intero, & ombreggiato, *Perspectiva Pictorum et Architectorum*, vol. I, fig. 47

5 The scenes of heaven

Andrea Pozzo's experience in the sacred theatre of the Jesuit Colleges, such as that of Brera in Milan, and later in the Roman College and the Roman Seminary, and again in Rome, at the Chancellery, is documented in the treatise in numerous plates. These are the *Teatri scenici* – as Pozzo calls them – similar to those discussed above, but in which, he explains, 'it is more difficult to find the perspective point' ([6]). Pozzo also addresses, in the second part of the *Perspectiva*, the question of scenic *soffitti* or ceilings, in particular those made with *aria con nuvole*, 'air with clouds'.

In the churches, the practice of covered roofs, flat ceilings and rounded apse, vaults and domes, lent themselves to the most audacious inventions and the most daring solutions, in keeping with a trend that had established itself in the course of the 1600s in churches and palaces. In this camp as well Pozzo was an innovator, taking the scenographic devices already tried many times and pushing them to new heights, not contenting himself with opening the ceiling to a vision of the heavens, but erecting a second platform of architectonic scenery over the nave, which then in its turn opens itself to the sky. He thus crowns the church with a palace, and crowns the palace with a depiction of heaven in a vortex of illusionistic foreshortening whose vanishing point coincides with the figure of the saint ascending into heaven. This occurs in the false cupola created in the church in Mondovì (Fig. 10) where he reduplicates the decoration of the nave and opens loggias for angels playing musical instruments. He then perfects his art in the very famous vault of the church of Sant'Ignazio in Rome. As noted, there are also numerous false architectonic cupolas painted on canvas, thanks to which Andrea Pozzo can creates illusions and astonishes the faithful, while sparing the coffers of the Jesuit order.

Fig. 10 Mondovì, Chiesa della Missione, frescoes in the central vault showing St Francis Xavier in glory

6 The characters in the scenes

All of the architectonic scenes painted by Pozzo in vaults and domes and on canvases for apparatuses are alive with figures. In addition to allegorical, angelic and demonic characters, and in addition to the divine characters and the saints of the Jesuits to be glorified, the scenes are filled to the brim with lesser characters who climb, fall, remain in precarious equilibrium, wink and point.

Some seem to emerge from the paintings along with parts of the stucco, such as the figure of a man in the vault of Sant'Ignazio who runs in front of a window; as noted by Fagiolo dell'Arco, his arm is not painted, but is actually formed in stucco placed on the window ([8]). This is why some paintings in oil show the interior of a church and a rich perspective scene populated with figures (Fig. 11). The great apparatuses for the forty-hours devotion, shown earlier, are also abound in characters, always on a miniscule scale with respect the majestic spatial volumes, tiny figures that daringly climb balusters and balconies, as though defying the empty space below.

The self-portrait of Andrea Pozzo conserved in the Chiesa del Gesù in Rome (Fig. 12), shows the artist in a similar pose, seated daringly on a cornice in the church Sant'Ignazio with one hand firmly holding (so as not to fall) onto the books about perspective that he learned from, and with the other, a finger raised, pointing to the vault and the cupola still to be frescoed.

Ingenious scenic apparatuses, daring compositions of canvases, or dizzying perspective solutions that are *sott'in su*, 'bottom to top', constitute the complex technical foundations of the visions that provide spectators with the materialisation of a principle of faith, the immediate perception of an ineffable truth that involves the senses and predisposes the spirit to meditation and prayer.

Fig. 11 A. Pozzo and assistants, *Ultima cena*, 1708, oil on canvas, Trento, Museo Diocesano

Fig. 12 A. Pozzo, Self-portrait, oil on canvas,
Rome, Chiesa del Gesù

References

1. F.S. Baldinucci, Vita del padre Pozzo gesuita (ms., 1725-1730 ca). In: A. Matteoli (ed.), Vite di artisti dei secoli XVII-XVIII: prima edizione integrale del Codice Palatino 565. De Luca, Roma, 1975.
2. A. Battisti (ed.), Andrea Pozzo. Luni, Milan-Trento, 1996.
3. E. Bianchi, D. Cattoi, G. Dardanello, F. Frangi (eds.), Andrea Pozzo (1642-1709) pittore e prospettico in Italia settentrionale. Temi, Trento, 2009.
4. R. Bösel, L. Salviucci Insolera (eds.), Mirabili disinganni. Andrea Pozzo (Trento 1642 - Vienna 1709) pittore e architetto gesuita. Artemide, Roma, 2010.
5. F. Camerota, Il teatro delle idee: prospettiva e scienze matematiche nel Seicento. In: R. Bösel, L. Salviucci Insolera (eds.), Mirabili disinganni. Andrea Pozzo (Trento 1642 - Vienna 1709) pittore e architetto gesuita. Artemide, Roma, 2010, pp. 25-36.
6. S. Carandini, Dalle quinte del teatro alla macchina d'altare. Andrea Pozzo e le pratiche sceniche del suo tempo. In: H.W. Pfeiffer, S.J. (ed.), Andrea Pozzo a Mondovì. Jaca Book, Milano, 2010, pp. 156-179.
7. V. De Feo, V. Martinelli (eds.), Andrea Pozzo. Electa, Milano, 1996.
8. M. Fagiolo dell'Arco, Tra barocco e illuminismo. Andrea Pozzo, fratello Gesuita, pittore. In: S. Carandini (ed.), Chiarezza verosimiglianza. La fine del dramma barocco. Bulzoni, Roma, 1997, pp. 193-217 .
9. B. Kerber, Andrea Pozzo. Walter de Gruyter, Berlin-New York, 1971.

10. K. Noheles, Teatri per le Quarant'ore e altari barocchi. In: M. Fagiolo, M. L. Madonna (eds.), Barocco romano e barocco italiano. Il teatro, l'effimero, l'allegoria. Gangemi, Roma 1985, pp. 88-89.
11. H.W. Pfeiffer, S.J. (ed.), Andrea Pozzo a Mondovì. Jaca Book, Milano, 2010.
12. A. Pozzo, Perspectiva Pictorum et Architectorum. Rome, Giacomo Komarek, 1693-1700. Anastatic rpt., A. Franceschini, L. Giacomelli, M. Hausbergher and A. Tomasi (eds.), Tipografia editrice. Temi, Trento, 2009.
13. S. Weil, The Devotion of Forty Hours and Roman Baroque Illusions. Journal of the Warburg and Courtauld Institutes XXXVII, 218-248, 1974.

Andrea Pozzo: Art, Culture and Mathematics

Marco Costamagna

Andrea Pozzo was born in Trento to a Milanese father, Jacopo, and his wife Lucia on 30 November 1642. He attended the Jesuit School in Trento until he was 17, but because his results that were less than brilliant, his father apprenticed him to a painter, given that booklearning went against his nature, and he preferred drawing and doodling to studying. From an early age he worked, and learned the art of painting in painters' workshops in Trento and Venice. It is thought that Andrea became practiced in this art and in the use of perspective forms through painting scenes for the theatre, every aspect of which was evolving in that particular period. In 1665 he chose the religious life, entering the *Society of Jesus*; although he never completed the entire *cursus* of Jesuit studies, he would remain in the order as a temporal coadjutator. What little is known about his life comes from the *catalogi breves* and the *catalogi triennales*, reports sent by the Jesuit fathers in the provinces fathers to the Father General at the beginning/ end of each year in the case of the former, every three years in the case of the latter. All of the evaluations between 1670 and 1705 concur in expressing great appreciation of his artistic talents. Over the years, his fame and his talent would both grow, but he remained simple and unaffected in his modest position as a *Brother* in the *Society of Jesus*. Andrea Pozzo died in Vienna on 31 August 1709. He is considered, along with Gian Lorenzo Bernini and Pieter Paul Rubens, to be one of the three great artists who expressed the art of the *Society of Jesus*.

Andrea Pozzo was a Jesuit, and this fact must be underlined in order to understand completely him as an artist. As mentioned, he is recognised as one of the three great artists who created Jesuit art, but he, more than anyone else, was also capable of representing philosophy, narrating history, and promoting the work of the Jesuits; in short, he expressed religiousness in a way that was absolutely extraordinary and avant-garde for the time. But at the same time, the *Society of Jesus* constituted a fitting environment for taking care of protecting this simple man, who was reserved, of few words, and of delicate health.

In order to understand fully who this great artist was, we must begin with the elements and characteristics of his technique, which are already evident in his first

Marco Costamagna
Restorer, Mondovì (Italy).

Emmer M. (Ed.): Imagine Math. Between Culture and Mathematics
DOI 10.1007/978-88-470-2427-4_6, © Springer-Verlag Italia 2012

work, the Jesuit church in Mondovì dedicated to St. Francis Xavier, one of the founding fathers of the Jesuit order.

Fig. 1 The Jesuit church in Mondovì (© Maurizio Roatta 2010)[1]

1 Historical background

The church in Mondovì dedicated to St. Francis Xavier was strongly desired by the Jesuits, who had by that time been carrying out their work in the city for over a hundred years. Construction work began in 1665, based on a design by Giovenale Boetto, architect to the house of Savoy.

At the beginning of the 1670s, a new rector for the Jesuit College in Mondovì arrived from Milan. By that point, work on the construction was well on its way; the shell had been completed, and the design by architect Boetto had been enriched with new details and trimmings to complete the work, but the Jesuits were not satisfied with the design: the part felt to be most faulty was the vault, which consisted of a poor excuse for a cupola in the first part of the nave, followed by a sequence of small vaults and sustaining arches that alternated in a disharmonic way up to the apse.

The Jesuits, in the person of the new rector, protested that the whole space was squat and unpleasant, and proposed creating a work that was more suitable for honouring both their great saint and the city of Mondovì, which at the time in Piedmont was second in importance only to Torino. This gave rise to heated debates which

---

[1] All the photographs of this paper are by Maurizio Roatta – Studio Roatta Architetti Associati Mondovì.

led to arguments between those involved: the designer, the master masons, and the Jesuits themselves. For their part, the Jesuits were even willing the demolish part of what had already been built in order to satisfy their aesthetic requirements, until finally, perhaps thanks to a proposal by the Jesuits themselves, the design was placed under the scrutiny of Andrea Pozzo, who was then at the Jesuit College in Brera, in Milan, in the capacity of a lay brother working as an assistant cook.

He was to maintain this status, if it can be called that, for a long time to come, but his skills as a painter, his genius, and his talent were already appreciated and recognised widely, even outside the order.

Pozzo studied the design and proposed several modifications. These included the addition of six columns in the nave that played no structural or load-bearing role; he wanted all the columns to be in faux marble; he also asked that two new windows be opened in the apse, and that the windows already present be widened.

As far as the vault was concerned – the critical point, object of all the arguments – he said in substance that he would deal with it himself with his paintings, and that 'all that is unpleasing now will become pleasing, and where there is now a flat ceiling, I will make a cupola'.

The modifications suggested by Pozzo were not enormous, but they completely changed how the architectural corpus was perceived. His idea was simple: he intended to create, by means of the real and the painted architecture, a sequence of scenic backdrops that would immerge the spectator in a unified, theatrical environment, and evoke his emotions by means of *special* effects, focussing all his attention on the apparatus of the altar.

Today his intentions are clear to all, but at the time I don't believe that even the Jesuits were completely aware of what Pozzo actually had in mind. But they had great faith in him; especially the new rector of the College of Mondovì, who knew him well. He had been there when this assistant cook had set up the scenic apparatus for the canonisation of St. Francis Borgia in the church of San Fedele in Milan, and in the church of SS. Andrea e Ambrogio in Genoa, and he had seen with his own eyes how successful these sacred theatres had been

Andrea's proposals were accepted, and he was commissioned to paint the vaults. He arrived in Mondovì in the spring of 1676.

(© Maurizio Roatta 2010)

The previous picture shows the famous false cupola photographed from the observation point, where the entire area comprised by the octagon is actually a flat plane. This image makes it possible to understand why Andrea Pozzo had wanted the columns in faux marble; he depicts identical columns in the paintings and thus gives the illusion of continuity with the real architecture, creating this effect of depth

Pozzo's painterly technique is called anamorphosis. The painting appears to us in its correct three-dimensionality only when observed from a precise point, a point which, in some works, the painter himself indicates for us by making a mark in the pavement of the nave. This is the first great work executed according to the canons of this technique, which for the time can be considered modern painting. Andrea Pozzo called the observation point 'the divine light'.

Many meanings have been attributed to that point. The Lutheran Christoph Friedrich Nicolai (1733-1811), one of Pozzo's fiercest critics, in 1781, shortly after the suppression of the *Society of Jesus* [in 1773], called it the point of blind obbedien*ce*, in denigration of the entire Jesuit order. But that point is the fulcrum of Pozzo's genius as a painter; it is the essence of his technique.

(© Maurizio Roatta 2010)

In 2009, while restoration work was still underway, the financial patrons sponsored an event on the occasion of the three-hundredth anniversary of the death of Andrea Pozzo, making it possible for visitors to use the scaffolding in place for the work to view the paintings close up. The images that show the *painter's point of view relative to the false cupola* allow us to formulate some comments about this technique.

(© Maurizio Roatta 2010)

Many visitors, upon seeing these strange tangles, after losing themselves a bit in contemplation, have asked in one way or another the same question:

'How did the painter do all this?'

A few answers have been attempted by experts both real and improvised, from the most improbable ('Andrea Pozzi left holes in the scaffolding so that he could see the painting from below and correct any errors') to the most ennobling and flattering ('Pozzo was pure talent. He was a genius'). This last assertion derives from the fact that although the painter in question was never a brilliant student during the years he attended the Jesuit College in Trento, he had already shown an early preference for painting over study; thus maintaining that Pozzo was a genius elegantly closes the discussion.

It is my conviction, without intending to take anything at all away from the artist's talent, that genius in this case is also supported by precise mathematical knowledge, and I think that Pozzo acquired this foundation precisely during his student years at the Jesuit College in Trento, and that he built on and used it for the rest of his life.

The church at Mondovì is his first work, and there may have been room for improvisation, but I am convinced that it is articulated by means of a graphic design that required precise calculations for the subdivision of the space and the correct distribution of the volumes. I believe that his system of calculating, if it can be defined in that way, has yet to be revealed. In order to execute this illusory cupola, he laid out a grid on the ceiling; traces are still evident in the plaster, and the nails are still there that were used to hold the strings to define the modules of a grid onto which was transferred with identical proportions a drawing created earlier. All of this indicates that behind the painting there was a precise design.

The figure that follows shows that Andrea Pozzo had no absolute need to go down below to check for possible errors. Here we are on the vertical axis of the point of observation, very close to the vault. As you can see, the part of the painted architecture is in the correct arrangement, and in spite of the fact that the central scene is still quite flattened, the cupola already begins to appear. Naturally, as you go down along this vertical line, that phenomenon becomes increasingly evident, finally assuming the correct dimensions.

(© Maurizio Roatta 2010)

(© Maurizio Roatta 2010)

The layers of plaster make it possible for us to see how he proceeded.

(© Maurizio Roatta 2010)

(© Maurizio Roatta 2010)

First he created these two lunettes in order to have the precise direction of the longitudinal axis of the church.

(© Maurizio Roatta 2010)

He then painted the scene of the ascension of the Saint, in the centre of the false cupola, very far from the centre of the actual ceiling.

(© Maurizio Roatta 2010)

He then laid out and painted the entire lower order of the architecture.

(© Maurizio Roatta 2010)

Next he painted the upper orders of the architecture.

(© Maurizio Roatta 2010)

Finally, he painted the figures.

(© Maurizio Roatta 2010)

In addition to calculating the perspective, there are also plays of interpenetration used by the painter to augment the effects, for example, the figure of the angel with the cello which appears to be suspended between the real space of the church and the heavens of Paradise. This of course is all fictive, as are the plays of light and shadow,

(© Maurizio Roatta 2010)

but fiction imitates reality in this scenic apparatus, since it represents a miracle: the levitations attributed to St. Francis Xavier, described in a Papal Bull.

(© Maurizio Roatta 2010)

We are now in the winter between 1676 and 1677. The apse has been completed with paintings, and Marco Mutis has erected the first four columns in plaster and completed the walls. With the help of three assistants and a team of carpenters and joiners, who are charged with erecting the scaffolding and constructing the various frameworks on which the canvasses will be stretched, Andrea Pozzo creates the scenic apparatus that was to serve for both the commemoration of St. Francis and for the celebrations of the *Quarant'hore*, or forty hours' devotion, and for other particularly important church functions.

Pozzo was a man of the theatre and kept the group working at a fast pace. Careful observation shows parts of the wood left unvarnished and hastily fitted joints. The painter himself seems to have used something more akin to a broom than a paintbrush, and to have painted directly on the canvas without having first primed it. Often the canvasses are not large enough to cover the entire wooden framework; in that case, as the figure shows, the painter painted directly on the wood.

Pozzo's work in Mondovì is ephemeral, constructed with simple materials, the stuff of theatre, designed to be dismantled and remounted. There may even have been an intention to rebuild it in marble in order to make the church even more

(© Maurizio Roatta 2010)

sumptuous, but as fate would have it, it has remained as it was for more than three hundred years. We restorers, who have had the opportunity to study every inch of this apparatus at length, have found no signs to indicate that it has ever been dismantled.

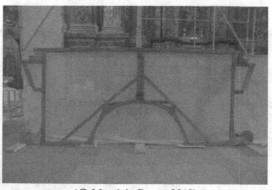

(© Maurizio Roatta 2010)

Fig. 2 Design for an altar apparatus similar to that in Mondovì (image taken from vol. I of the treatise *Prospettiva dei Pittori e architetti* by Andrea Pozzo, Rome, 1780)

"I have used this tabernacle several times for the exposition of the forty hours devotion" said Andrea Pozzo with regard to the design which is shown in fig. 61 of his treatise. This provides us with a first precise reference for the analysis of the perspective construction of the apparatus in Mondovì. The second reference is the physical model, which is the high altar of the church of Santi Giovanni e Paolo in Venice (a work in marble by Mattia Carneri and Baldassarre Longhena). The altar apparatus in Mondovì is a very similar copy, differing only in the ornamental figures that crown the tympanum and in the materials. It is composed of two parallel

backdrops, each of which is composed of various fitted panels, ten for the front wing, and four for the back one. Together they provide the image of an architecture that is coherent with that of the scenic apparatus of the church, not only in the proportions but also in the illusions of materials and the use of the architectural orders. The scenic apparatus is supported on planking at the top of the high altar.

The images that follow provide an idea of how this apparatus functions, based on the remaining elements that constitute the mechanism. A simple apparatus, the working parts consist in a few pulleys, a pulley knot, and a small winch placed in the upper part of the ceiling of the apse.

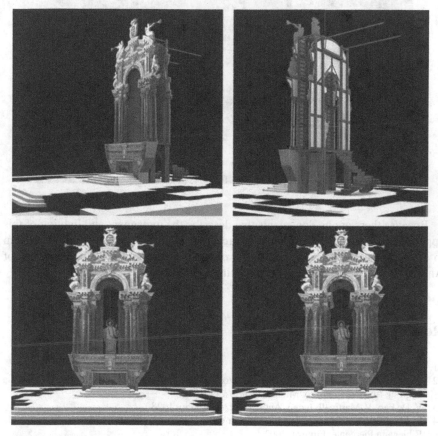

Fig. 3 Details of the two backdrops and of the structure of the apparatus

It is presumed that to create an evocative scenic effect, the church was darkened while the two backdrops of the apparatus were illuminated with artfully located oil lamps. Marks left by nails indicate that cloth panels arranged in the central space of the first backdrop acted as drop curtains. At a certain point of the function, the drop curtains opened, St. Francis appeared, and the faithful watched him levitate.

Because there are no documents that describe the movement and lighting of this apparatus, a group of scholars has been created to study and apply the correct system to a scale model constructed ad hoc in order to then apply it to the apparatus of Andrea Pozzo, which was and is unique in the world.

References

1. B. Kerber, Andrea Pozzo. Berlin, 1971.
2. W. Gramatowski, Il profilo di Andrea Pozzo alla luce dell'archivio romano della Compagnia di Gesù. In: Andrea Pozzo. Milano-Trento, 1996.
3. V. De Feo, V. Martinelli (eds.), Andrea Pozzo. Milano, 1998.
4. E. Filippi, L'arte della prospettiva. L'opera e l'insegnamento di Andrea Pozzo e Ferdinando Bibiena in Piemonte. Firenze, 2002.
5. H.W. Pfeiffer, Andrea Pozzo a Mondovì. Milano, 2010.

Homage to Hypatia

Hypatia as Polymath

Michael A.B. Deakin

1 Introduction

First let me introduce my subject and explain my title.

Hypatia was a mathematician, astronomer and philosopher whose life spanned the late IV and early V centuries AD. She was, for a time, the world's leading mathematician, the only woman ever for whom this claim can be made. She is also the first female mathematician of whose life we have any reasonably complete account (although we have sketchy details of a number of earlier figures). I have published a biography [1] of her, which contains English translations of all the primary sources from which we draw the story of her life, and because these sources constitute the basis for detailed claims I advance in what follows, I here refrain from reproducing the minutiae of the relevant documentation, but instead refer the reader to the place in my book where these details are supplied.

She was the daughter of Theon of Alexandria, himself an astronomer and mathematician, best remembered as the principal source of our knowledge of Euclid's *Elements* [2]. The date of her birth is uncertain, and various authors have argued for different years ranging from 350 to 375 AD. The best modern scholarship opts for ca. 355 AD, but with a large error bar attached. She lived out her life in Alexandria, which was at that time the intellectual hub of the Eastern Roman Empire, surpassing then both Athens and Constantinople in this respect. She died, brutally murdered, in (almost certainly) 415 AD, a victim of an outpouring of Christian fanaticism against a prominent adherent of a rival philosophy. Her womanhood, her learning and her violent martyrdom have combined to make her a powerful symbol of scholarship under difficult circumstances and an icon for the feminist movement.

The term *polymath* applies to one expert in a variety of different fields of intellectual endeavor. Its etymology is Greek: "poly", of course, meaning "many", and "math" signifying "learning". This latter half of the word, however, requires further elaboration. Its origins go back to the Pythagorean movement (VI century BC).

Michael A.B. Deakin
Mathematics Department, Monash University, Clayton (Australia).

Emmer M. (Ed.): Imagine Math. Between Culture and Mathematics
DOI 10.1007/978-88-470-2427-4_7, © Springer-Verlag Italia 2012

At their gatherings, there were two classes of participants: the *mathematikoi*, who were authorized to contribute to the discussion, and the *akousmatikoi*, who could attend and listen, but not themselves play an active part [3] (think, in this connection, of the word *acoustic*). Because Pythagoras is recognized as a pioneering mathematician (in our modern sense of the term), the element *math* in the first of these designations has come to be applied to that specific branch of learning that today we call *mathematics*.

I find this particularly apt for my theme here. While it is clear that Hypatia's expertise covered several fields (which makes her a *polymath* in the usual general sense), it is also clear, as I will argue in detail, that, of these various fields, it was the mathematics that was paramount. In other words, the second linguistic element in the word *polymath* may be assigned a particular significance. We might rephrase the meaning of the entire word as "many-fold ramifications of mathematics".

It is especially important to stress the primary rôle played by mathematics in her thinking, because many, indeed most, earlier biographies either gloss over her mathematical endeavors or else discuss them quite inadequately. For critical details on this matter, the reader is once again referred to my book [1].

2 Hypatia as Mathematician

The various branches of the mathematical sciences were later combined and formalized into the *quadrivium* ([1], p. 178) of the medieval universities. As this name implies, there were four such branches: arithmetic, geometry, astronomy and music. Hypatia was adept in the first three of these, and may also have had some involvement with the fourth. (Music was seen as mathematical ever since the time of Pythagoras and his interest in the mathematical principles behind, e.g., the monochord. However, this aspect of the story will not be pursued here. Furthermore, we nowadays discuss astronomy as a separate discipline for reasons I outline below; I will follow this convention here.)

But now first note that we *must* think of Hypatia as being first and foremost a mathematician. The books she is credited with writing are *all* concerned with mathematics or astronomy; no others are mentioned. Her strictly mathematical works comprise commentaries on the *Conics* of Apollonios and the *Arithmetic* of Diophantos.

Her times were not, however, propitious for the pursuit of mathematics. Mathematics had become confused with numerology and astronomy with astrology. Both of these were regarded with mistrust by the Christian establishment of the late Roman Empire. The thirty-sixth canon of the council of Laodicea forbade priests to be mathematicians. Hypatia's almost exact contemporary St. Augustine of Hippo recalled his adventures with *mathematicians* thus ([1], p. 64):

> Those imposters whom they call *mathematicians* I consulted without scruple: because they seemed to use no sacrifice nor pray to any spirit for their divinations: which art, however, Christian piety consistently rejects and condemns.

It follows that the interest Theon and Hypatia displayed in mathematical endeavors, although actually pursued only in its reputable aspects, could easily be misrepresented and accordingly mistrusted. Indeed there is very good evidence that such misrepresentation was used to justify Hypatia's murder ([1], pp. 148-149).

In such a climate, the priority for both father and daughter was not so much the advancement of the frontiers of knowledge (i.e. research mathematics), but rather the urgent attempt to preserve and transmit the body of existing knowledge. This was done in two ways, first by the production of commentaries on mathematical and astronomical classics, and second by teaching, hoping to keep alive the flame of genuine learning in the living beings of their students.

Theon had worked extensively on the books of Euclid and Ptolemy, especially the former's *Elements* and the latter's *Almagest*; Hypatia extended his program to the more difficult works of Apollonios and Diophantos, producing commentaries on both the *Conics* (Apollonios) and the *Arithmetic* (Diophantos). These commentaries are both now presumed lost, although it is possible that parts of them may be preserved as interpolations and translations in later works. Again, the matter is discussed in more detail in my book ([1], Chapter 9).

I have already stressed that it is the *mathematical and astronomical works* that are listed in the primary source material. No specifically philosophical treatise has ever been ascribed to her. She is, however, also described there as giving public lectures on philosophy, and we may plausibly deduce the general nature of the philosophy she espoused, a task I embark on below.

This mathematical aspect of her thinking even attracted adverse comment. The philosopher Damaskios compared her unfavorably with his own teacher, Isidoros ([1], pp. 140-143):

> There was a very great difference between Isidoros and Hypatia, not simply because she was merely a woman while he was a man, but also insofar as she was expert mainly in geometry whereas he was a true philosopher.

(Women and mathematicians alike will rejoice in this reaction by the French historian of mathematics, Paul Tannery: "In plain language, Isidoros knew no mathematics!"; see [1], p. 177.)

3 Hypatia as Astronomer

Although in her own time, Hypatia's interest in astronomy would have been seen as being essentially mathematical in character, I here follow modern nomenclature in separating astronomy from mathematics proper (arithmetic and geometry deal with abstract mental constructs, whereas astronomy has not, and indeed cannot, shed its material referents).

Hypatia is known to have collaborated with her father in at least one aspect of his work on Ptolemy. Quite what was the nature of her contribution to the third book of her father's commentary on Ptolemy's *Almagest* is under dispute. It is also unclear whether or not this contribution is the same as an *astronomical table* she is credited with producing.

However, it has been plausibly suggested that her contribution was an improved method for the *long division* algorithms needed for astronomical computation (see [1], pp. 115-118). In Book III of the *Almagest*, there is a division calculation aimed at computing the number of the degrees swept out by the sun in a single day as it orbits the earth (remember that the Ptolemaic system of astronomy was geocentric). In modern terms, the determination of the value of 360/365.25. The computation proceeds by means of a tabular method that is superior to the approaches used in other parts of Theon's commentary. It may be that this tabular method is what the sources refer to as the *astronomical table*, but this is by no means the only possible interpretation.

Hypatia also assisted her pupil Synesios in the design of an astrolabe that he had had made as a gift to an influential official. The astrolabe in question would have been a *little astrolabe*, which used a stereographic projection of the celestial sphere to represent the heavens on a plane surface. The underlying theory seems to have been passed down from Ptolemy, or perhaps even the earlier Hipparchos via Ptolemy, to Theon and thence to Hypatia and Synesios.

In his covering letter, Synesios wrote ([1], p. 60):

> Astronomy itself is a venerable science, and might become a stepping stone to something more august, a science which I think is a convenient passage to mystical theology, for the happy body of heaven has matter underneath it, and its motion has seemed to the leaders in philosophy to be an imitation of mind. It proceeds to its demonstrations in no uncertain way, for it uses as its servants geometry and arithmetic, which it would not be improper to call a fixed standard of truth.

The passage bears elaboration. The composition of the heavens was seen as involving the *quinta essentia*, a fifth *element* distinct from the other four: the mundane earth, air, fire and water (ordinary matter). Synesios saw astronomy as a bridge between the material world and a more sublime counterpart, visible in the sky. This counterpart is directly influenced by the abstractions of geometry and arithmetic, which provide guaranteed truth (he is citing Ptolemy at this point). Elsewhere in his writing, Synesios saw dreams as a similar *bridge* between the material and the spiritual realms.

The scribe Photios, summarizing an earlier account of Hypatia, wrote ([1], p. 158):

> [Philostorgios] says that Hypatia the daughter of Theon was taught mathematics by her father, but reached an excellence far above her teacher, especially in astronomy, and that she instructed many [pupils] in mathematical studies.

(As indicated above, in accordance with the custom of the time, he regarded astronomy as a branch of mathematics.)

4 Hypatia as Philosopher

The primary sources tell us that Hypatia was a Neoplatonist, which means that she would have ascribed a religious cast to the philosophy of Plato. This label of itself is not particularly specific, and could be applied to a wide multiplicity of actual beliefs. However, Synesios of Cyrene was her devoted pupil and *his* position is very well documented. Not only did he speak most fulsomely *of* her, and indeed in his letters *to* her, but he also sent her copies of some of his writings, in effect inviting her to referee them. She must have approved, because the works remain extant.

Synesios idolized Hypatia (see [1], pp. 150-158). He addressed her as "mother, sister, teacher and withal benefactress". He referred to her as "the lady who legitimately presides over the mysteries of philosophy" and spoke of her as providing Egypt with "fruitful wisdom". We are thus led to think that Synesios' philosophical outlook was close to Hypatia's own, and thus that we may attempt to reconstruct Hypatia's philosophy by extrapolating from Synesios'.

Synesios was very much a Platonist and so followed Plato in espousing a theory according to which the world of everyday experience is actually a projection of a more fundamental reality – a universe of *forms*. We can access this deeper reality via our power of abstraction and the clearest example of this principle at work is the case of mathematics.

Indeed so important is the example of mathematics in Plato's philosophy that he is credited with the injunction: "Let no one ignorant of mathematics enter [my academy]."

So familiar do mathematical concepts become to us that we reify them, and think of numbers such as 2 as *actually existing* – not of course in a concrete sense, but nonetheless as being every bit as *real* as concrete objects. So, for example, the number 2 is reached via our abstracting from pairs of objects (hands, feet, parents and the like). But beyond this simple example, we even apply this same mode of thought to much more elaborate cases such as the number π which indeed is referred to as a *real number*!

The concepts of mathematics can thus be regarded as clear examples of *Platonic forms* – probably the clearest examples there are. It is most certainly true that most working mathematicians adopt a form of Platonism, perhaps even subconsciously, but not necessarily so, as a working philosophy. Here is an explicit statement from the mathematician Charles Hermite:

> There exists, if I am not mistaken, an entire world which is the totality of mathematical truths, to which we have access only with our mind, just as the world of physical reality exists, the one like the other independent of ourselves, both of divine creation [4].

The attribution of *divine creation* requires elaboration. In 1856, Hermite embraced Roman Catholicism and clung to it rigorously. His attribution of a divine locus to mathematical reality is thus hardly surprising (see, in this connection, my remarks below on the close correspondence of the notions of the Christian God and the Neoplatonic One). However, by no means all mathematicians who ascribe an *otherworldly* realm to mathematical reality follow Hermite in his religious interpretation.

Fig. 1 Portrait of Hypatia

Another mathematician, an avowed atheist, G. H. Hardy has written in quite similar vein, but naturally without the reference to the godhead:

> I believe that mathematical reality lies outside us, that our function is to discover or observe it, and that the theorems which we prove, and which we describe grandiloquently as our *creations*, are simply the notes of our observations ([5], pp. 123-124).

Even those mathematicians who decry *platonic mathematics*, such as Philip Davis, nonetheless succeed in demonstrating that it accurately describes the way in which most working mathematicians think [6].

The case against *platonic mathematics* has also been argued forcibly by Lakoff and Núñez, who nonetheless acknowledge its force ([7], p. 80).

> The metaphor *Numbers Are Things in the World* has deep consequences. The first is the widespread view of mathematical Platonism. If objects are real entities out there in the universe, then understanding Numbers metaphorically as Things in the World leads to the metaphorical conclusion that numbers have an objective existence as real entities out there as part of the universe. This is a metaphorical inference from one of our most basic unconscious metaphors. We barely notice it.

They continue in this vein and derive three consequences of such a view, the first of which I find tendentious, and will refrain from discussing further. But their second reads "[n]umbers should not be product of minds, any more than trees or rocks or stars are products of minds" and their third "[m]athematical truths are discovered, not created" (when, in the above, they say "numbers", they imply also reference to other mathematical constructs).

Their second *consequence* leads to an important philosophical division as to the nature of mathematical *reality* Lakoff and Núñez espouse the antithetical view: that mathematical constructs are the product of (embodied) human minds. This view impinges on the ongoing debate as to whether mathematical advances are *discoveries* or *inventions* (their third *consequence*). It is not my intention to involve myself in this debate; but I do I point out, indeed stress, that, for many mathematicians, the Platonic account is paramount, and even for those who dispute it, there is recognition and acknowledgement of its force.

The Platonic view extends naturally into mathematical education:

> ... the exact sciences [are not] based on an accumulation of statistics. In order to teach the young that three plus four makes seven, you do not add four cakes plus three cakes nor four bishops plus three bishops nor four cooperatives plus three cooperatives nor four patent leather buttons with three wool socks. Once the principle has been intuited, the youthful mathematician grasps that three plus four invariably make seven and he does not have to prove it over and over again with chocolates, man-eating tigers, oysters or telescopes ([8], pp. 123- 124).

An explicitly religious dimension can be given to the theory of forms by supposing a further act of abstraction, directed to discovering an even deeper reality underlying the forms themselves. This deeper reality was termed the One or the Unity. For Neoplatonists, the goal of the well-spent life was seen as one of conformity with the principles implicit in the One. The virtuous life entailed a quest for mystic union with the One.

This concept of One is actually not very different in its fundamentals from the God of the monotheistic religions: a fundamental correspondence clearly grasped by Synesios, who converted to Christianity without compromising his fundamental Neoplatonism. However, it was equally clearly *not* grasped by the Christian fanatics who assassinated Hypatia.

In fact, there are Neoplatonist ideas implicit in many of Christianity's doctrinal formulations. In particular, the doctrine of the Trinity has clear Neoplatonic roots, especially in its notion of the *Logos* or Word, a term derived from Neoplatonism, but routinely applied to *God the Son* who in Christian belief became incarnate as Jesus of Nazareth (such a view clearly characterizes the opening sentences of the Gospel of John).

It may well be true that Hypatia's Christian contemporaries could see the idea of mathematics as a path to the divine as perhaps odd or quirky (although clearly Synesios did not), but there is nothing inimical to Christianity in such a view. Rather, a Christian would see Neoplatonism as incomplete, in that it omitted reference to the Incarnation, the central dogma of Christianity.

The compatibility of Hypatia's outlook with Judaeo-Christian theism is most strikingly illustrated by the fact that she was not acted against during the episcopate of Theophilos (bishop of Alexandria from 385-412 AD). Theophilos had sacked the pagan temple of Serapis and replaced it with a Christian church dedicated to St. John the Baptist, of whom he held custody of some alleged relics. But this ultra-militant Christian took no action against Hypatia and indeed enjoyed a cordial friendship with her pupil Synesios. With the deaths of Synesios and Theophilos, and the accession of St Cyril of Alexandria to the bishopric, Hypatia lost two powerful protectors. I think it probable that Theophilos took the view that, although Hypatia was not herself a Christian, her philosophy was nonetheless not anti-Christian (as other streams of Neoplatonism in fact were), and so she was not an antagonist of the Christian position, but perhaps rather even a somewhat distant ally.

The form of Neoplatonism that Hypatia adopted formed the basis of a devoted lifestyle, aimed at communion with the One. In her case, it led her to embrace a strict celibacy, as a corollary of an attempt to rise above material things. It did not however mean that she withdrew from public life. Indeed, her pupil Synesios urged in his *Dion* that such engagement was to be encouraged (see [9], Chapter VI). Her lectures on philosophical topics were popular and well-attended. Furthermore she engaged in civic affairs and held frequent discussion with the prefect of Alexandria, Orestes. This aspect of her life was, sadly, a part of the motivation for her murder, for when a Christian assassination attempt on Orestes failed, its perpetrators chose a softer target: Hypatia.

5 Conclusion: the Primacy of Mathematics in Hypatia's Philosophy

So I here argue that Hypatia made the example of mathematics her primary entree to the world of Neoplatonic philosophy. Indeed, this approach earned her some disapproval, as instanced by the comment by Damaskios (quoted above).

In the first place, the example of mathematics offers the clearest possible illustration of the Platonic theory of forms, so that someone committed to that would necessarily make mathematics an integral part of their philosophy, as indeed Plato himself did.

Moreover, the truths of mathematics stand scrutiny without qualification, and without reference to what we might call *experiment*. If we like, we may describe them as necessary facts, rather than contingent ones. In such a sense, we may therefore view them as offering a path to aspects of the One, as the fundamental underlying reality.

We therefore see Hypatia as unifying the mathematics of her day into an overall Neoplatonic philosophical outlook: the mathematics was a *stepping stone* to higher things; it enabled access to "a fixed standard of truth".

So she *did* use mathematics as a gateway to philosophy, and thus was not a *mere geometer*, as Damaskios would have it; unlike Isidoros, she was a true philosopher, who knew, treasured and used mathematics.

References

1. M.A.B. Deakin, Hypatia of Alexandria, Mathematician and Martyr. Prometheus Press, Amherst, 2007.
2. G.J. Toomer, Article on Theon of Alexandria. Dictionary of Scientific Biography, Vol. 13, C. C. Gillispie et al. (ed.). Charles Scribner's Sons, New York, 1976, 321-325.

3. P. Gorman, Pythagoras, A Life. Routledge Kegan Paul, London and Boston, 1979.
4. C. Hermite, Quoted in G. Darboux' La vie et l'oeuvre de Charles Hermite. Revue du mois 1(46), 1906; English translation by K. M. Lenzen, The Mathematical Intelligencer 5(4), 13.
5. G.H. Hardy, A Mathematician's Apology. Cambridge University Press, Cambridge, 1967.
6. P.J. Davis, Fidelity in mathematical discourse. Is one and one really two?. American Mathematical Monthly **79**, 252–263, 1972.
7. G. Lakoff, R.E. Núñez, Where Mathematics comes from: How the Embodied Mind brings Mathematics into Being. Basic Books, New York, 2000.
8. J.L. Borges, A. Bioy-Casares, Chronicles of Bustos Domecq. Dutton, New York, 1976.
9. J. Bregman, Synesius of Cyrene, Philosopher-Bishop. University of California Press, Berkeley, Los Angeles and London, 1982.

Women's Contributions to the Progress of Mathematics: Lights and Shadows

Elisabetta Strickland

It's undoubtedly worthwhile to analyse the role played by women in mathematics. Men have obtained many recognitions in this field, but the same cannot be said about women. Indeed, how many people are aware of the contributions of Hypatia, Émilie du Châtelet, Maria Gaetana Agnesi, Sophie Germaine, Mary Fairfax Somerville, Sonya Kovalevsky and Emmy Noether? Nevertheless, today we can confirm that these women made substantial contributions to the progress of mathematics. For this reason they deserve our attention, but also because they had extraordinary lives and peculiar personalities which are interesting to observe closely.

It has been proven that when human beings started to develop the concept of number, women approached this concept in the same way as men [1, p. 11]. As a matter of fact, anthropologists are convinced that primitive women had a relatively high order of creative intelligence, absolutely as lively as that of primitive men.

Cuneiform documents on clay tablets from the region of ancient Mesopotamia have shown that around 4700 B.C. there were already expert mathematicians at work and when the Code of Hammurabi was in force among the Babylonians, women took care of business and accounting, exactly as in ancient Egypt. Nevertheless, none of the women experts in mathematics in these ancient cultures are known by name: we have to wait for the Hellenic age to discover the names of women who were expert in mathematics.

The school of Pythagoras, which started around 539 B.C. in the Greek colony of Croton in southern Italy, was also attended by women, and some of the teachers were women too. Pythagoras married one of them, Theano, and two of their daughters ran the school after the death of Pythagoras.

The first notable woman mathematician was the Egyptian Hypatia, born around 370 A.D. in Alexandria [2]. She was the daughter of the mathematician Theon, who was the librarian of the Library of Alexandria and educated her as if she were a boy. In about 400 A.D., she became headmistress of the Platonist school at Alexandria, where she imparted the knowledge passed down from Plato and Aristotle. Hypatia was a popular lecturer, drawing students from all parts of the empire. She was a Hellenistic pagan. Her contributions to science are many: she is credited with the

Elisabetta Strickland
Department of Mathematics, University Tor Vergata, Rome,
Gender Interuniversity Observatory, Sapienza University of Rome (Italy).

Emmer M. (Ed.): Imagine Math. Between Culture and Mathematics
DOI 10.1007/978-88-470-2427-4_8, © Springer-Verlag Italia 2012

invention of an astrolabe [1, p. 28], wrote commentaries on Diophantine equations and on the conics of Apollonius, edited her father's commentary on Euclid's *Elements*, wrote a text called *The Astronomical Canon*, contributed to the invention of the hydrometer and the hydroscope, and worked on the charting of celestial bodies. When Bishop Cyril and his Christian followers accused her of causing religious turmoil, Hypatia was assassinated in a ferocious way: a Christian mob of monks waylaid Hypatia's chariot as she travelled through the town. The monks stripped her naked, dragged her through the streets to a church, where they killed her, flaying her body with sharp oyster shells and burned the parts. This cruel death didn't prevent Hypatia from securing her place in history.

After the fall of Constantinople, there was a decline in intellectual progress; even after the Renaissance the status of women changed very slowly. One has to wait until the seventeenth century to hear again of a prominent woman mathematician.

In France during the Age of Reason it was not easy for women to have an education which was not too superficial. The only institution which performed this duty was founded during the reign of Louis XIV, the *Sun King*: the Institut of Saint-Cyr, where young ladies were prepared to become the wives of aristocrats.

This was the general situation of education in France when in 1706 Émilie de Breteuil was born in Paris, daughter of Louis Nicholas le Tonnier, baron of Breteuil, head of the protocol at Court [1, p. 52]. She had very good teachers at home and they were all astonished by her capability in understanding mathematics. Often they didn't even understand properly what she was saying, as her ideas were much more sophisticated than those they were accustomed to. When she was nineteen years old, she married the Marquis of Châtelet, who was colonel of a regiment. Émilie du Châtelet was left often alone and spent her time studying and enjoying society. She was so smart in mathematics that all the best minds in Paris became her friends. She fell in love with one of them, Voltaire, and went to live with him in the country house of the Châtelet family, at Cirey-sur-Blaise. There du Chatelet was introduced to the work of Leibniz and Newton, which occupied her mind for fifteen years. She became so deep in her thoughts, that she inspired Voltaire to write the novel *Candide*. The couple could live their love story freely because du Chatelet's husband admired Voltaire and decided to tolerate their *menage a trois*. Du Chatelet liked her routine: she could work every morning on her papers and in the evenings she could devote herself to the pleasures of the fashionable world, which suited her personality as much as science books. During those years, she published with Voltaire the *Istitutions de physique*, which was devoted to the work of Leibniz, and translated Newton's *Principia* from Latin, establishing her reputation for competence among contemporaries.

When she was expecting her fourth child, she admitted that she didn't know who the father was, maybe even not Voltaire, as at the time she was having an affair with the poet Saint Lambert. She really didn't care about the issue and preferred to spend those months studying Newton's theories. Voltaire himself wrote that when the baby was born, the little girl was put on a volume of geometry and the mother went to bed taking the papers she was writing along with her. Apparently du Chatelet seemed well, she spent her convalescence together with her husband, Voltaire and Saint

Lambert taking care of her. But all of a sudden, late one afternoon, she died quietly. Voltaire was beside himself, and declared that he had lost not only his lover, but half of himself.

While these peculiar events where taking place in France, in Italy women were becoming interested in the study of mathematics. The most notable among them was Maria Gaetana Agnesi, born in Milan on May 16, 1718.

Agnesi was a child prodigy; she learned seven languages before her thirteenth birthday, and her father, Pietro Agnesi, who held a chair at the University of Bologna, and mother, Anna Brivio, gave her the best possible education, with special emphasis on mathematics and philosophy, so that she learned from the work of masters such as Newton, Leibniz, Fermat, Descartes and Euler. Her father organized seminars at home and distinguished intellectuals gathered to listen Agnesi's lectures; but she was shy by nature, and asked her father for permission to enter a convent in order to spend her life studying in peace. This was not granted, not least because Agnesi had twenty-one siblings and half-siblings and when he lost his last wife, Agnesi was asked to help with their education. Her first book, which she started writing at the age of twenty, entitled *Instituzioni analitiche ad uso della gioventù italiana* (*Analytical Institutions for the use of Italian Youths*), was a treatise on differential and integral calculus. She spent ten years on it and when it was published in 1748, the academic world was extremely impressed by her work. But her reputation reached extraordinary heights when she discussed a planar cubic curve, originally studied by Fermat, which she called a *versiera*, from the Latin word *vertere*, as the curve actually turned. But this word was interpreted as an abbreviation of the word *avversiera*, or 'wife of the devil', so it happened that John Colson, professor of mathematics at Cambridge, translated the word *versiera* as 'witch', and the curve discussed by Agnesi became known as the *witch of Agnesi*.

Because of her results in mathematics, Agnesi became a member of the Accademia delle Scienze of Bologna and was given a diamond ring by the Empress Maria Theresa, to whom she had dedicated her book. Even Pope Benedict XIV showed appreciation for her work and secured a teaching position for her at the University of Bologna after her father's death. But Agnesi preferred to devote herself to help the poor, and she was appointed head of the female section of the Hospice Trivulzio in Milan. She died at age 81: the city of Milan named a street after her, and a crater on Venus is named in her honour.

Like Émilie du Châtelet, another notable mathematician, physicist and philosopher of the eighteenth century was also French: Sophie Germaine, born in Paris on April 1, 1776.

Germaine was 13 when the Bastille fell, and as the turmoil in Paris made it impossible for her to spend time outside the house, she turned to her father's library. In the book of Jean-Étienne Montucla, *Histoire des Mathèmatiques*, she read the story of the death of Archimedes, killed by a Roman soldier during the sack of Syracuse in 212 B.C. while he was totally absorbed by a problem in geometry. Germaine decided that if mathematics had held such fascination for Archimedes, it was a subject worthy of study. The family didn't approve of her attraction to mathematics and, when night came, they would deny her warm clothes and proper lighting in her bedroom. She waited until they went to sleep, took out candles, wrapped herself in blankets and worked until dawn. One morning Germaine was found asleep on her desk, the ink frozen in the ink horn; this made them realise that she had to be free to do mathematics [1, p. 85].

In 1794, when she was 18, the École Polytechnique opened, but as a woman Germaine was not permitted to attend. She managed to obtain the lecture notes of Lagrange, who was a faculty member, and started to send him her work using the pseudonym of M. Le Blanc. Lagrange recognised the quality of her work and was determined to meet M. Le Blanc, so Germaine was forced to disclose her true identity. Lagrange didn't mind that Germaine was a woman, and helped her in her work.

Initially Germaine had a deep interest in number theory and, again under the pseudonym of M. Le Blanc, she wrote to Carl Friedrich Gauss, presenting some of her work on *Fermat's Last Theorem*. Gauss thought well of Germaine, even after he discovered that she was a woman, but he generally did not review her work, so finally the correspondence ended without the two having ever met.

In 1811 Germaine participated in a competition sponsored by the Académie des sciences in Paris concerning the experiments of the physicist Ernst Chladni with vibrating metal plates. The object of the competition was to give a mathematical theory for the vibration of an elastic surface. She didn't succeed this first time, because the jury said the equations were not established, but she tried again and at the third attempt, in 1816, she won the prize and this allowed her to have a reputation on a par with that of Cauchy, Ampere, Navier, Poisson and Fourier.

But Germaine's best work was in number theory, as she made a significant contribution to *Fermat's Last Theorem*. She wrote again to Gauss to ask his opinion on her proof for a special case, but Gauss never answered. As we know, Andrew Wiles solved the problem in 1994, but Germaine's work two hundred years before is considered substantial.

· Finally Gauss did his best to convince the University of Göttingen to confer a doctor's degree *honoris causa* on Germaine, but she never had the chance to receive it, since she died of breast cancer in 1831. When the state official who had to fill her death certificate came to her house, he refused to write 'mathematician' for her profession, writing *property owner* instead. When the Eiffel Tower was erected and it was decided to inscribe in it the names of the seventy-two scientists whose work on elasticity contributed to the enterprise, Germaine's name was left out, perhaps because she was a woman. But today Germaine is considered one of the founders of mathematical physics.

We now leave France and turn to the British contribution of women in mathematics. This leads us to some interesting reflections on the life of Mary Fairfax Somerville, whose extraordinary mathematical talent came to the surface by a lucky accident. She was born in Scotland, at Jedburgh, in 1780. When she was a little girl, she didn't like studying; as a matter of fact at the age of ten she was barely capable of reading. One day, when she was a teenager, she was leafing through a fashion magazine and noticed some algebraic symbols, which she found fascinating [3, p. 46]. She was at that time attending Nasmyth's Academy, in order to learn painting and dancing and there she overheard the master of the school talking about Euclid's *Elements of Geometry* and succeeded in obtaining a copy of the book through her youngest brother's tutor, together with a text of algebra [3, p. 49]. She started to spend her time studying at night at candlelight, behaviour which her family found deeply worrying. It was decided that she would marry her cousin Samuel Greig. This is exactly what happened, and in the first three years of marriage she had two sons, one of whom died shortly after birth. Her husband was against her wish to study mathematics, but he died in 1807, leaving her financially independent for the first time in her life. By this time she had pretty well mastered geometry and astronomy.

At the age of 33, she won a prize solving a problem on Diophantine equations which had been posed by a popular science journal. The editor gave her a list of classics in mathematics suited to form a solid background in the field. Her family still found her manner of life abnormal and convinced her to marry a second time with another cousin, William Somerville. This time she was lucky, as her new husband was an intelligent and handsome surgeon, who moved with Fairfax Somerville to London, where she had the chance to study and meet all the most outstanding intellectuals of that time, such as the mathematician Pierre Laplace and the explorers Georges Cuvier and Sir Edward Perry.

In 1826 Fairfax Somerville submitted an article to the Royal Society on the magnetic properties of ultraviolet rays in the solar spectrum and in 1838 she was appointed honorary member of the Royal Astronomical Society, together with Caroline Lucretia Herschel. Her second husband also had health problems and Fairfax Somerville had to move first to Paris, then to Italy. This didn't prevent her from writing two volumes on behalf of the Society of Useful Knowledge, *Mechanism of the Heavens* about the work of Laplace and one about Newton's *Principia*, which had to be translated from Latin into English for the British academic institutions. Notwithstanding her role as mother of five children (she had four from the second marriage), in 1843 her book *The Connection of the Physical Sciences* was published and was followed in 1848 by another one called *Physical Geography*. At the age of 89, after the death of her second husband, she wrote *Molecular and Microscopic Science*, a treatise on the form and the rotation of the earth and the tides of the ocean and atmosphere, plus other texts on various topics. She also wrote also her memoirs, *Personal Recollections* [3]. She died at 92, in Naples. Most of the popularity of her writings is due to her clear style and her enthusiasm for the subjects. No one has any doubt about her being one of the most outstanding British women scientists.

Up to know, we have talked about women mathematicians in France, Italy and Great Britain. What can be said about other countries?

An interesting answer can be found in Russia. Sofia (or Sophie) Vasilyevna Kovalevskaya was born in Moscow in 1850. Her father, Vasily Vasilyevich Korvin-Krukovsky, served in the Imperial Russian Army. She was the first major Russian female mathematician as well as a public advocate of feminism and a noted writer whose works include both fiction and nonfiction.

There are some transliterations of her name; she herself changed it in Sonya Kovalevsky in her last academic publications. Her mother Yelizaveta decided to nurture her interest in mathematics by hiring a well known tutor for her higher education,

Alexander Nikolayevich Strannoliubsky, who taught her calculus [4, p. 7]. Because her father prevented her from completing her education in Russia, she decided to contract a *fictitious marriage* with Vladimir Kovalevsky, a young palaeontology student and together they emigrated from Russia. Kovalevskaya began attending the University of Heidelberg, in Germany, studying under such teachers as the physicists Helmholtz and Kirkhoff and the chemist Bunsen [5, p. 424]. But when she learned that the mathematician Karl Weierstrass was teaching in Berlin, she asked him to give her private lessons: the University would not allow her to audit classes. In 1874 she presented three papers to the University of Göttingen as her doctoral dissertation: the first was on partial differential equations, containing the well known Cauchy-Kovalevsky theorem, the second one on the dynamics of Saturn's rings and the third one on elliptic integrals. Weierstrass supported her and she obtained her doctorate in mathematics, summa cum laude, without having to pass oral exams or defend her thesis, thus becoming the first woman in Europe to hold that degree. After this achievement, she returned to Russia and gave birth to a daughter, Fufa. After a year devoted to raising her baby, she resumed her work in mathematics: in order to do this, she left her husband, who suffered from mood swings, and entrusted Fufa to her sister, Anyuta. Vladimir couldn't stand this and, also beleaguered by a stock swindle, committed suicide.

Kovalevskaya was distraught over this tragedy, and asked the Swedish mathematician Gösta Mittag-Leffler, whom she had known as a fellow student of Weierstrass, to help her. She thus obtained a position as a docent at Stockholm University. In 1884 she was appointed a professor (without a chair) and became the editor of *Acta Mathematica*. In 1888 she won the Prix Bordin of the French Académie des sciences for the celebrated discovery of what is known as the 'Kovalevsky top', on the complete integrability of a rigid body motion about a fixed point: the only other tops were those of Euler and Lagrange. Finally in 1889 she was appointed Professor with a chair, the first woman to hold such a position in northern Europe. She also became a member of the Russian Academy of Sciences, but was never offered a professorship in Russia. She died of complications from a flu in 1891 at the age of forty-one, after a trip of pleasure. She was buried in Sweden, at Solna. The Soviet Union honoured her with a postage stamp showing her portrait. Kovalevskaya also wrote a memoir, *A Russian Childhood* [4], plays (in collaboration with Anne Charlotte Edgren-Leffler, the sister of Gösta Mittag-Leffler), and a partly autobiographical novel, *Nihilist Girl*.

At this point it's a duty to talk about Emmy Noether, another extraordinary mathematician, the founder of modern algebra. She was born in a Jewish family in Germany, at Erlangen, in 1882. Her father, Max Noether, professor of mathematics at the University of Erlangen, was at that time already a celebrity, thanks to his theory of algebraic functions. Noether had a wonderful character, she was full of joy and loved having fun, so she grew up dividing her time between social activities and her father's lectures, which were attended also by her brother Fritz. She obtained her degree in mathematics in 1907, defending a thesis 'On complete systems of invariants for ternary biquadratic forms' under the supervision of Paul Gordan. She herself defined her work 'a jungle of formulas'. After her father's death, David Hilbert convinced Noether to go to Göttingen, where he believed she could be useful, with her knowledge of invariants, for his joint work with Felix Klein on the general theory of relativity. Hilbert tried desperately to convince the university to give her a position, but he didn't succeed, because she was a woman. He was so exasperated by the refusal, that one day at a faculty meeting he burst out with his famous words 'This is a university, not a bathhouse!'. But Hilbert had such an high opinion of the young lady, that he ignored the faculty's decision and started to send her to give lectures in his place. Only at the end of the first world war was Noether officially allowed to teach at the university. In the 1920s she developed the theory of ideals in commutative rings. She loved working at the beautiful Institute of Mathematics of Göttingen, which had been constructed with the financial help of the Rockefeller Foundation.

In 1932 she was the first woman to give a talk at the International Congress of Mathematicians, in Zurich. But under the Nazis, because she was Jewish, she was compelled to leave for the United States, where she was offered a job at *Bryn Mawr College* in Pennsylvania, very close to the Institute for Advanced Study of Princeton. During her American years, she worked on noncommutative rings and hypercomplex algebras and united the theory of group representations with the theory of modules and ideals. But she didn't enjoy the privileges she had received in the States for long, as she died at 53 as the consequence of an apparently successful operation. Albert Einstein wrote in the *New York Times* [May 4, 1935, p. 12] that 'Fraulein Noether was the most significant creative mathematical genius thus far produced since the higher education of women began'.

Hilbert [1, p. 152] said that in Göttingen people usually referred to Noether as 'der Noether', i.e., using the masculine article, because of a respectful recognition of her power as a creative thinker who appeared to have broken through the barrier of sex.

One last glimpse at Italy: as the 150th anniversary of its founding as a modern state is being celebrated at the moment of this writing, it's quite natural to wonder if women have made important contributions to mathematics during this period.

Let me mention two relevant names in this context: Maria Pastori (1895-1975) and Maria Cibrario Cinquini (1905-1992). Pastori had as master and colleague the great mathematical physicist Bruno Finzi. She didn't come from a wealthy family or grow up in an intellectual environment, but nevertheless she succeeded in entering the Scuola Normale Superiore di Pisa, where she defended her degree thesis in 1920. In 1939 she obtained a chair at the University of Messina and in 1947 she returned to her hometown Milan, holding the chair of rational mechanics. During her career she published about one hundred papers, obtaining important results in tensor analysis and relativity. She wrote a book in collaboration with Bruno Finzi, entitled *Calcolo Tensoriale ed Applicazioni* (*Tensor Calculus and its Applications*), still considered today a milestone in mathematical literature.

Maria Cibrario Cinquini obtained her degree in 1927 at the University of Turin, under the supervision of Guido Fubini. As an assistant professor she worked with Giuseppe Peano and in one of her first eight papers written during the six years after her graduation, she analyzed 24 statements derived from the definition of limit, following Peano's work. After his death, she worked with Francesco Giacomo Tricomi and Guido Fubini, and won the Corrado Segre Prize for young assistant professors three years in a row (1926-1928). She became internationally renowned for her discovery that hyperbolic-elliptic differential equations could give a description of transonic aerodynamic phenomena. She married her colleague Silvio Cinquini, and they had three children. Her family duties didn't prevent her from solving the Goursat problem for nonlinear hyperbolic equations and the Cauchy problem for systems of first order differential equations.

Some final words about women in mathematics today. Many women are first-class mathematicians: as a matter of fact, women have never before enjoyed such prominence in mathematics. At the most recent International Congress of Mathematics, many of the 178 speakers were women, including Parimala, an Indian mathematician, Claire Voisin, French, winner of the Clay Research Award in 2008, and an Italian, Matilde Marcolli, who works at Caltech in California.

Moreover, in 2010 for the first time a woman was elected President of the International Mathematical Union: she is the Belgian Ingrid Daubechies, first woman full professor of mathematics at the University of Princeton, expert in the theory of wavelets. And again in 2010 a woman was elected President of the European Mathematical Society: Marta Sanz-Solè, Spanish, from the University of Barcelona, an expert in stochastic processes.

No woman has yet been the recipient of the Fields Medal, the most important prize awarded for mathematics, but we have good reason to hope that in 2014, when the next International Congress of Mathematicians takes place in Seoul, Korea, the prize will finally be awarded to a woman.

References

1. L.M. Osen, Women in Mathematics. Cambridge, MA, MIT Press, 1974.
2. M. Dzielska, Hypatia of Alexandria. Cambridge, MA, Harvard University Press, 1996.
3. M. Somerville, Personal Recollections. From Early Life to Old Age. London, John Murray, 1874. Available at: http://www.gutenberg.org/ebooks/27747. Last accessed 19 June 2011.
4. S. Kovalevskaya, A Russian Childhood. Heidelberg, Springer, 1978.
5. E. Temple Bell, Men of Mathematics (1937). New York, Touchstone, 1986.

Hypatia's Dream

Massimo Vincenzi

1 The plot

The play narrates Hypatia's last day. From when she wakes up in the morning, then leaves home for school, until her assault and death. The narration is alternated with the recollection of one of the protagonist's *desperate* feats: saving the library of Alexandria. A feat that we have turned into a paradigm of her entire life. This recollection is alternated with the more and more vehement and violent voice of the political and religious authorities. Beginning from Theodosius' first edict in 380 AD and culminating with Bishop Cyril's anathemas. Most of the narration concerning Hypatia, though faithful to the historical documents, has been freely reinvented. The text relating to the political authorities is taken from the four Theodosius' edicts. For the part of the narration concerning Bishop Cyril we have freely readapted fragments of his speeches, using as guidelines the available historical evidence.

2 The story

If reason and faith represent the two parallel tracks along which Western history has travelled during the past two thousand years, the episode that better typifies these two contending ideologies took place in March 415 with the murder of Hypatia (Alexandria of Egypt, circa 370 – 415 AD), known as *the muse* or *the philosopher*.

The historical context in which the event took place is the time when Christianity underwent a genetic mutation when it stopped being persecuted after Constantine's edict in 313, became a State religion with Theodosius' edict in 380 and in 392 engaged in persecution, in its turn, when the Greek temples were destroyed and the *pagan* books burned.

Massimo Vincenzi
Teatro Belli, Rome (Italy).

Emmer M. (Ed.): Imagine Math. Between Culture and Mathematics
DOI 10.1007/978-88-470-2427-4_9, © Springer-Verlag Italia 2012

The events in Alexandria came to a head starting from 412, when the fundamentalist Cyril was appointed patriarch (proclaimed Saint and Doctor of the Church in 1882).

In just three years, availing himself of an armed wing of fighting monks, he spread panic in the city. But his true sacrificial victim was Hypatia, the most famous cultural figure in the city. Daughter of Theon, rector of the University of Alexandria and himself a famous mathematician, Hypatia and her father made scientific history for commenting the Greek classics: we owe the editions of the works of Euclid, Archimedes and Diophantus to them.

In a world that, to this day, is still almost exclusively a men's world, Hypatia is remembered as the first female mathematician in history: the equivalent of Sappho for poetry or Aspasia for philosophy. Indeed, she was the only female mathematician for more than a millennium: for others to appear, we have to wait until the eighteenth century. But Hypatia was also the inventor of the astrolabe, the planisphere and the hydroscope, as well as the leading Alexandrian exponent of the Neoplatonic school.

Her works have gone missing. Anything we know about her comes from the letters of Synesius of Cyrene, her favourite student.

Hypatia's rationalism, who never married a man because she said she was already "married to the truth", was a far too conspicuous counterpart to Cyril's fanaticism.

One of them had to yield and it could be none but Hypatia. Hypatia was attacked in the street, her body flayed with ostraca, dismembered and then burned. Governor Orestes reported the event to Rome, but Cyril claimed that she was safe and sound in Athens. After an inquiry, the case was dismissed "for lack of witnesses".

After centuries of guilty silence, in these past few months, Hypatia's figure is forcefully coming to the fore. Radio and television programmes, articles in the leading newspapers, numerous successful books have also been published and at the festival of Cannes a Spanish MovieTalk has been presented.

3 Hypatia'S Dream

(Hypatia's house, dawn)

Hypatia: I love the wind that blows through the trees in my garden. It makes a strange sound. I love to hear the voice of the trees. They speak to me. They remind me of the stars. The rustling of the sky when at night I look at the lights above. It is the music that keeps me company. The same prolonged, sweet note wishing me good night that I hear on waking up in the morning.

Enough! I am captivated by a thousand thoughts and forget what is important. My father used to tell me again and again. I've tried to escape his arguments in whatever way, but at the end logic, his logic followed me everywhere. He was right. Study first and foremost. If I'm here today, it's thanks to him, to my father, it's thanks to studying. To his passion. If I'm here today.

I could laugh if I were not in despair. Where am I today? Where are we all today?

I must get up. Enough! Enough going around with words. I must see my pupils. I must go to them. I've been home too many days. I must go out. I want the sun on my face. I want to wear my nicest dress. To dismiss evil thoughts. They want to keep us home. They want the field clear. Nothing. No witnesses.

What kind of God fears words. Hates books. It's a strange God, indeed.

But it's the God they wish to portray that is scary. The projection of their fears. It's not God. Because you can call God or the gods with many names, but if they are really up there, beyond the stars and they govern everything, why should they be afraid? And of what? Of us? Fear and violence belong to man, to God, if there is one, belongs love.

Off-stage voice: The august emperor Theodosius to Albinus, praetorian prefect. No one shall violate one's purity with sacrificial rites, no one shall immolate innocent victims, no one shall go near sanctuaries, enter temples and turn one's gaze towards statues sculptured by mortal hands so as not to deserve divine and human sanctions. If anyone who is devoted to profane rites enters the temple with the intention of praying, he shall be immediately forced to pay the sanction with public demonstration.

(1ˢᵗ flashback)

Hypatia: Let's go, don't be afraid. It's they who don't understand. Who still don't understand. Hurry with those chests. They are books. Many books. As many as they've never seen. An army of books. We shall take them home. Our soldiers. They will be our weapons. You'll see the look on their faces when our house shall be full of books again. They ordered to burn them, the books of the Alexandria library, but they shall not find them if we rescue them. Cheer up! The stars shall soon rise to give us a hand. Come on, all together. You, on the boat deck. You, down here to load the horses. One on the lookout on the hill to check if anyone approaches. Let's hurry. We must unload before daybreak.

Off-stage voice: Those who have betrayed the holy faith and desecrated holy Baptism shall be banned from common society: they shall be exempted from testifying in court, they shall have no part in wills, shall inherit nothing and shall not be named heirs by anyone.

For those who have corrupted the faith, the disgrace of morality shall not be effaced by penance, which shall only be good for the other crimes.

(Hypatia's house, dawn)

Hypatia: Two more minutes. I'm tired. Too many nights I slept an unrestful sleep. I shudder at every noise. My sight is getting worse. That's how old age must be. I don't even know whether I'm old anymore. The calendar has gone mad. The sundial shows a shadow that is impossible to decipher. I would cry if I knew how. I dreamt of my father. I dream of him more and more often. We used to spend hours with our heads bent over the same sheets of paper. We always whispered. How many words I didn't understand as a child. I would stretch my muscles until my arms and legs ached. I stretched out in the effort. I would become taller, very tall. Then I would collapse in a heap exhausted. When I opened my eyes, my father would be there, staring at me. Smiling. I dreamt that they hammered down our instruments, my astrolabe, my hydroscope. Piece by piece. Clubs to smash everything. I dreamt that they burned our books. I know this will really happen. I cannot predict the future, I just line up logic. That's why I must go. My pupils are waiting for me.

Off-stage voice: No one shall be granted the authority to perform sacrificial rites, no one shall wander around the temples, no one shall turn one's gaze towards the sanctuaries. In particular, all profane entrances shall be identified, which remain closed hindering our law so that, if something should instigate anyone to ignore these bans concerning gods and sacred things, the transgressor shall understand that he shall be stripped of any indulgence.

(2nd flashback)

Hypatia: Where are you? Where are you? Here you are, I feared I had lost you.

Hurry! The stars are high in the sky. There is light. We can see because we can follow the stars. They can't see anything. They are small. In their small houses. Their small palaces that look like huts shaken by the wind of their very fear. If there is a God, he is not above them. If there is a God, he is above us all. Check that all the books have been brought down, on the wharf. Don't leave in the waves the smallest piece of paper. If their God existed, his hands would be black with ink. His eyes red from reading. He would have read all the books in the world. Their God. If he existed. He would have room for all the books of Alexandria. If he existed.

I promise you. They will not burn these books. They will not be banished from Alexandria. Remember that. This city will die when the last book shall be lost. This city will hold out to everything. Even to them. Even to their God, if we'll manage to save even one page of our past, of our knowledge. This city will not die. I promise you.

Off-stage voice: No one, of any gender, order, social class or status or honorary role, either of noble birth or humble origins, in any place, no matter how far, in any city, shall sculpture simulacrums or offer any innocent victim to the gods or secretly burn a sacrifice to lars, genies or penates, light fires, offer incenses, or lay wreaths to these idols. For, if it shall be heard that someone has offered a sacrificial victim or has consulted the entrails, shall be accused of lese-majesty and shall accept the proper sentence.

(morning, Hypatia leaves home)

Hypatia: The way is long. I read as I walk. As if the book were not an obstacle, but a kind of compass. I count the pages and know exactly where I am. Three pages and my garden is behind me. Ten pages and my old wet-nurse's house. How many mornings I spent with her. It's comforting that she died before all this happened. It's even a joy.

My dear child, she used to say to me. Your father is a man of great intelligence and a good man. But he's always a man and he doesn't know what a child, what a woman needs. In your world, always leave room for your heart. We women are not weak. We are only different. And saying this she laughed.

Even among the stars there are different stars, female stars. I learned this growing up. I've seen stars with a paler light, with a slimmer body. Stars that always seemed on the verge of dying out. Always about to be pushed down from the sky by other bigger stars. Instead, the years go by, the space wind blows to wipe out the starlight, but they're still here. And they always look more beautiful, year after year. And they will be there forever. Like us, Like me, I read. I read and walk. I read their texts. To understand what they write. But I don't understand. What has happened? They spit their insults in our faces. They have created a wall of hate. Today they call it church. What will they call it tomorrow? They imagined a warrior god and decided it was their god. Good and just. This is what saddens me the most. The deception of thought, of minds. Faith and hope soiled by fear. Paralyzed by terror. God is not only yours. I feel like screaming. God is not like this. I would like to scream. But screams don't belong to me. I don't like the sound of iron clashing against iron, the sounds of battle. I would like to stop and talk to them. To explain, to explain myself. To open up our minds. I've done so a thousand times with my pupils. Many times. At first, the staring eyes of those who do not understand, then the spark of doubt and finally the happy smile of those who have gone a step forward.

I read and I think I see them. I see them on the white walls where they write their insults. I see them behind the windows that shut down as I go by. In the people who bump into me on purpose. Increasingly violent. I see them as soon as I raise my head from my papers. I no longer know whom I should fear.

I feel there is less people who follow me. What has happened? Once, even those who governed the city came to me for advice, my opinion had bearing on their decisions. And now? There is no one around me. My pupils lie to me: they say that today one is ill, that maybe another will come tomorrow. They tell me not to worry, that things will change. They can't persecute us. They can't hurt us. They experienced pain when they were forced by the Romans to pray in the Catacombs. Will they do that to us? This is what my pupils say. But they are young and idealistic.

They don't know how dangerous can be a god generated by resentment.

I see a shadow. I jerk my head up and look around me. I just have to get to my school. Find my pupils. Think. But I can no longer think.

I see them. Now I see them. I hear their voices and I'm not dreaming. I walk faster. No! I stop.

I don't want to gratify them. I stop and go through the pages of my book with deliberation.

I hear them, though. They say that I'm impudent. That I am mad. That's the course they'll follow. Madness. I already know. The light on the walls will still be the same, the shadows won't be too long when I will become the village madwoman. The heretic. A new word they invented. Books with long, endless lists of heretics: that's what the future holds in store for us.

Off-stage voice: To be accused of a crime, it shall in fact suffice the will to oppose the same law, to pursue illegal actions, to manifest occult things, to attempt to do forbidden things, to seek a salvation other than Christian salvation, to promise a different hope.

(3rd flashback)

Hypatia: It's a quiet night and the sky is full of stars. On a night like this, my father took me on the roof of our house to look at them. He never stopped searching for the supreme order that eluded him. If they only knew that he was the first to speak to me of God. But it was a God that united instead of dividing. He told me that planets do not revolve at random following their instincts like animals in the forest. He often used animals as a metaphor so that I would understand. He knew I liked tales. So he used to tell me about stars and plants, giving them the names of the animals in our garden. Thus the planets were no longer mad animals, but they all obeyed to a higher order. At the time we didn't know what order it was. And I doubt I know it now. But that was a wonderful thing. I can't explain it. I only remember feeling a sense of peace. I would often crouch on the roof alone and stare at the stars. I also wanted to search.

Search, look, understand.

Enough talk! Let's go, the city is still asleep. Let's not wake those who must not be awakened. This is not a night to be taken lightly. It's a night that bites.

Off-stage voice: If anyone has venerated mortal works and worldly effigies with incense and, ludicrous example, fears even those that they represent, or has erected altars with dug lumps of earth to the vain images, that he be accused of slander to the full religion (Christian) and be guilty of violating religion. That he be fined in his household things and belongings, having become a slave of pagan superstition. Then, all the places where sacrifices have been offered with incense be seized.

(morning, on the street)

Hypatia: The few friends I meet almost pretend not to see me. If they speak to me, they say: stay home, we will come to see you. Don't you see them? Don't you know they hate you? Of course I know. Of course I see them. That is why I go out. I want to look them in the eyes. I want to chase them. I am the hunter. I was not born to end up on an altar, the lamb of a sacrifice I do not recognize. Then, my nightmare comes back to my mind. It's been hounding me for many nights. Do you know what a 1 by 2 meter cell dug behind the wall of a church is like? Do you know what is like being always in the dark? In a stone cage that spreads itself on you? This is my nightmare.

I am locked up in my fear, a prisoner, worse than being buried. I don't move. I fall on my knees and feel no pain. A trickle of blood awakens me for a moment. But it's only a dream, I tell myself. Then my father appears. My pupils are shadows, they're not here with me. I see their men everywhere. Shadows behind the windows. Monks. The bishop's men. They shout "God is with us!". I am scared. I feel the iron inside my flesh. I must awake from this dream. From this terrible nightmare. Calm down, calm down, I tell myself. But I'm mute.

Then I wake up.

I walk towards the school. A shadow slips behind me and hastens to the opposite side of the street. A cart hides it from me. I probably imagined it. I run a hand over my eyes. I reopen the book I had closed. I shall soon be home. In my true home, my school, and before that my father's school and before that my father's teachers' school.

I shall soon be home.

Off-stage voice: I, Cyril, bishop of Alexandria, implore you: my Emperor, give me the land purged of the heretics and I shall repay you with the sky. Help me crush the heretics and I shall help you crush the Persians!

(4th flashback)

Hypatia: Wait. I am no longer a little girl. I am not like you. I cannot think of passing through the night on a white horse as if I were twenty years old. Fatigue weighs on me. But I was not always like this, you know. No, your teacher was young. I was young and -people say – also beautiful. Very beautiful, someone wrote to me. I never really believed it. Vanity is a sin only for those who are here now. Only a blind God can place it among sins. Vanity is just a sweet thought that women need to get through the darkest hours. Your teacher was also vain. And I pride myself. It's one of the sins they spit on me. If it is a sin, then yes, I am guilty. Do you hear what they say? They reproach me for going round the streets with you without shame. And what should I be ashamed of? Of being a woman?

Go! I'll stop here to rest a while. Yes. I know the way. It's my house we are going to. There we shall all find some peace. Myself, you and our books. Go, I shall only close my eyes for a moment.

With my eyes closed, I can see well what my pupils cannot even perceive. There shall be no feasts for us who saved Alexandria's books. We shall not walk through the triumphal arch. Our army shall not take up arms again. Our heads are bowed. Our faces desperate. Lit by the burning paper. Prisoners of ropes and chains. There shall not be dances for us. They shall not lay tables in celebration. Only the noise of their rhythmic steps. The obsessive sound of their prayers beating our breast. God is with us, God is with us, God is with us... I can't just sit here waiting for the end, watching thoughts die. I cannot.

Off-stage voice: God's church is constantly threatened by "heresies", by impure and unholy doctrines, by godless persons, full of inanity, excesses, boundless ignorance and depravation. These persons are highly unholy, they are slanderers and deceivers by right, they are undermined by the seeds of viciousness and seriously affected by ignorance of God. Their highest degree of stupidity and their folly leads them to profess diabolical doctrines. Their contempt of God shall plunge them into hell, if they haven't already died a horrible death in this life.

(at the school, afternoon)

Hypatia: No one. No one. There is no one in my school. I walked this far hoping to find someone to talk to. Not even to teach. Now, I would be satisfied to talk. But in these days of terror, my dearest friends are locked up in their houses. As if fear could save them. In the street, people laugh at the woman being beaten, at the woman they are calling witch. They don't know that soon it will happen to men with sparse hair. Then to fat men. Then to those whose skin is too dark or too fair. And then to misfits. The kingdom of lunatics awaits us. When men believe they are God. Towers become steeples. Temples become cathedrals. They will burn books. Then they will burn schools. Statues. And then they will burn us. I know. I know. I tried to shout it.

But my cries came back. The buffets of indifference.

I love this place. I love the books that are here. I grew up with them. Will memories suffice to protect me?

Off-stage voice: Then on with the bursting wave of these men... go on with the gossip and foolish prattle, with words embellished with chimeras and deception! Oh, God! Help me crush the heretics.

(5th flashback)

Hypatia: The first sun ray awakens me. Blinds me. I try to cover my eyes with my hand and I feel the tears. I have heartburn. My tears tell me to stop again. My body captures me in its old age.

I must leave. I must get up from this rock. I grab the horse's reins. The sun shows me the way. The air is pleasant. The books smell of dust and dew. They smell of happiness. No God could hate this smell. I see the seagulls playing with the waves. I feel the salt on my skin. I feel like touching the sand with my feet. The path is steep. I remember my father, when I used to persuade him to come here. I remember when, after jumping in the ocean, I turned my head in the waves, to see if he was waiting for me on the beach. And yes, he was there, waiting for me!

I want another dawn, just for me. The books are safe. I would like to think I'm at peace now.

Off-stage voice: If they do not convert themselves, the Lord shall shine his sword against them. They are at the extremity of viciousness. Their throat is indeed a wide open tomb... their lips conceal the viper's venom. Come to your senses, intoxicated people.

God is with us!

(at the school, evening)

Hypatia: Who are you? Who are you? I do not wish to see anyone. Leave me alone. They knock again. Go away. Go away. It's them. They want me to become a Christian. They offered me money for this. Money for my school. But I cannot let them buy me. If I let them buy me, I will no longer be free and be able to study. Religion, any religion, any dogma or ideology, if it doesn't allow you to think, becomes a cage that stifles you. I go round the empty benches. How many days I spent here with my pupils. They can't have abandoned me too. Again those noises. I fear everything. They drove me to this. I see shadows and hooded heads in the garden, where once there were trees. I hear stealthy footsteps where once there was the joyous running of my pupils after classes.

Who are you? Who are you? No. The hands knocking at my door are not friendly. I would like to shout that I'm not afraid. To say that we are many in here. And that we will defend ourselves.

But I'm alone and they know it. Who are you? Who are you?

Flames and screams! Flames and screams! Flames and screams!

I bend my head and crouch.

Tongues of flames chasing me.

I'm scared. But I'm not scared of God.

I'm scared of them.

I would confess all the sins in the world, if it would be of use.

If it could stop the hands and fists.

Blows. Kicks. Blows with stones cutting my skin.

My body thins out. I am transparent.

But I still want to exist.

Maybe it would suffice to open my eyes.

But I no longer have eyes.

They ripped them off.

Images spin. My girlish smile, my father's untroubled face, the laughing at school, the voices.

I repeat the names of the stars that I loved.

I think of the order of the universe.

The pages of my book burn with me. Then nothing more. Only flames. Flames and screams. The screams of their defeat.

Don't listen to what they'll tell you.

They want me to disappear in thin air.

That's why they'll say I left, I ran away.

That's why they'll say that no one saw what they did to me, no one heard.

Don't believe them!

They burn my body and my writings because they want nothing to be left of me.

But they're wrong.

Thought does not burn.

Remember that.

I see you, my friends.

I'm there with you. As you run fast clinging to your horses.

The night will take us to the sea.

Free.

The wind in the sails.

A ship full of books.

Words to the oars.

Numbers to the rudder.

The order in our heads giving us the course.

We shall find a place where we can start anew.

Another dawn awaits us.

And don't look back.

To watch my body burning.

Thought does not burn.

Now, all I want is to climb on my roof and look at the stars.

My father is up there, waiting for me.

My father is waiting for me.

I know.

4 A note by Carlo Emilio Lerici

I've been working with Massimo Vincenzi for many years. We became friends by chance in a pub in Trastevere in Rome where I used to spend my evenings after the theatre. The ideas for the plays came to us while chatting in the café. That's how BIRD IS ALIVE, a play dedicated to 6 great jazz musicians came about. Then the others followed: *Eyes to the sky*, the ghosts trilogy, *Alan Turing and the poisoned apple*, *Barney's version*, *Ruth Ellis* and *Hypatia's dream*. A word, a conversation, an interesting article, a book and we immediately got to work.

This play came about the same way, entirely by chance. In April 2009, *Alan Turing* was on stage and on the day of the last performance, when the play was already over, a person walked by the Teatro Belli and saw the poster. He became curious and came in to ask about the play. He was a member of an association named *Ipazia Preveggenza Tecnologica*. They were interested in Turing, the British mathematician, father of artificial intelligence. We met again and talked about many topics and projects. During one of these meetings, the association's director, Oreste Grani, told me about Hypatia. He was interesting in producing a play about her. They were ready to support it in every way. And that's exactly what they did. But they probably didn't expect us to get going so fast. After all, I had never heard of Hypatia before. Like many people, I completely ignored this amazing story. But when I spoke about it to Francesca Bianco, the actress I've been working with for more than 25 years, she was thrilled at the idea too. So we immediately called Massimo Vincenzi. He didn't know anything about her either, but he also became enthusiastic and set to work at once. Then we involved Francesco Verdinelli, the musician who wrote the score and with whom I have also worked for many years, and finally, I

Fig. 1 Poster of the play

called another long standing friend and co-worker, Stefano Molinari, whose voice we used to interpret the religious authority. Finally, Teresa Pedroni from the *Diritto & Rovescio* company, gave us her valuable help for the production of the play.

In September 2009, we were supposed to participate in *Opere Festival* in the Odescalchi Castle in Bracciano with the play *Assassins*, but we didn't want to waste the opportunity to propose to the festival directors Maurizio Conte and Alberto Bassetti to host *Hypatia's dream* as well. We asked him to provide us with an auditorium and we would offer him the play in exchange. A lunatic suggestion at a time when you're forced to assess everything in terms of money. But enthusiasm often drives people mad and so the directors, overwhelmed by our enthusiasm, even though the festival programme was already closed, accepted our proposal and provided us with a beautiful auditorium and an evening entirely for us.

After the *sold out* debut of September 19, we revived the play at the Antonio Salines' *Teatro Belli within the Theatres Festival*. It was October 4th and the play was scheduled for 10:30 p.m. Due to a series of technical problems, it started at 11:45 pm. The house was full. No one had left and no one had made a noise about the delay. Everybody wanted to see Hypatia.

That's why we decided to repeat the play on the following days. We always had a full house. *Hypatia's Dream* had been a success! We couldn't believe it. The newspapers wrote about the motion picture[1] that couldn't find a distribution, a book in Italian[2] was about to come out that would sell thousands of copies in a few days. Hypatia had compellingly come to the fore and we were there too.

Then the play returned on stage in November, in February and in March at the *Teatro Lo Spazio* in Rome, made available by Alberto Bassetti e Francesco Verdinelli who also took part in the production. Again, it had been a success. Afterwards, we were invited in Naples and in Genoa. The invitations are growing in number throughout Italy. The page we created for the play on Facebook has gathered more than 2000 enthusiasts in a short time. We found out that the motion picture was finally going to come out in Italy. I was looking for a theatre to stage the play again and, by a miracle, two plays that were on at the Teatro Belli had been cancelled. We could go back on stage!

I don't know whether this passion for Hypatia will continue after the release of the motion picture, but I truly believe that the wheels put in motion cannot be stopped. We certainly will not stop taking Hypatia across Italy.

For Hypatia.

---

[1] *Agorà*, directed by Alejandro Amenábar, Spain, 2009.

[2] A. Petta, A. Colavito, *Ipazia*, Vita e sogni di una scienziata del IV secolo. La Lepre, 2010.

Mathematics and Art

Modern Geometry versus Modern Architecture

Isabeau Birindelli and Renata Cedrone

1 Introduction: Hope

Too often mathematics and architecture are related through concepts that have been central in Renaissance, and through ideas that were new in the 15th century; in this research, we propose sophisticate geometrical ideas and shapes that are still today object of active researches in mathematics, such as curves, vector fiber bundle, fibrations, foliations, non-euclidean surfaces etc... in order to read architectural realizations as well as to "equate" objects with an artistic or design value. This is done independently of the awareness of the "conceiver".

We shall describe mathematical objects and their architectural counter part. The interest is two folded, on one hand we wish to raise philosophical questions such as: Is the artistic and architectural value of a building or object of design amplified or even due to its mathematical content? (*Philosophy is written in this grand book, the universe... It is written in the language of mathematics, and its characters are triangles, circles, and other geometric figures* Galileo []) or is mathematics only an instrument that can be used by the architects or the artists?

On the other hand, thanks to powerful algorithms, the mathematical objects have now an immediate way of being created or represented easily and quickly; at the same time "technical solutions" for complex architectures can be given thanks to new developments in building technic. Hence the imagination of the shapes to which the "architect" (or the artist) can be inspired, through the mathematical framework, goes well behind the simple "triangles, and circles" mentioned by Galileo. But of course some understanding is necessary.

To be more precise, we feel that we can easily affirm that any reader of this note, and most "educated" people know such artists as: Santiago Calatrava, Zaha Hadid,

Isabeau Birindelli
Sapienza University of Rome (Italy).
Renata Cedrone
Laboratorio di Progettazione e Pianificazione, Rome (Italy).

Emmer M. (Ed.): Imagine Math. Between Culture and Mathematics
DOI 10.1007/978-88-470-2427-4_10, © Springer-Verlag Italia 2012

Anish Kapoor, Pierre Boulez, Bruno Munari, and, somehow simplifying, we can say that their success is proportional to the wonder their creations generate.

Stratified manifolds, hyperbolic spaces, fibre bundles, riemanian surfaces, geodesics, Lorentz attractor, stochastic processes, homoclinic orbits, gaussian curvature, minimal surfaces; with the same certitude we feel that most people have never even heard of them or, if they have, they remain at best mysterious and at worst terribly boring mathematical concepts.

Now each of these complex concepts have been a source of wonder and excitement for whoever "invented" them, or even for every mathematician that has a moment of epiphany when she (finally) understand their "power". Often, their complexity, their richness can be glimpse at in their representation, while of course their "beauty" is much deeper then the esthetic one.

The artists mentioned above have sometime used the knowledge of these mathematical objects (even if sometimes that knowledge was only germinal). Of course this is not said in order to reduce their artistic value. The idea is just to use these examples of contamination between mathematics and architecture, to enrich the "language of the artist" and of the critic of architecture (which is in fact everyone of us who endures it). Precisely we hope to enrich everyone's vocabulary in order to enrich everyone's vision. A bad artist will stay a bad artist, no matter how much mathematics he learns, but a great artist will be potentially unpowered by a fantastic tool which is unknown to most, but is available now to many, thanks to technological supports.

2 Spiraling

We start with a very basic and classical parallel, in the hope of making you comfortable.

2.1 Curves

A very classical example of spiral used in architecture is the lituus whose equation is given in polar coordinates $r^2\theta = a$, and it is said to represent the volute of the Ionic column. But we don't want to dwell on this, being more interested in the symbolic content of the spiral.

One of the oldest architectural example of spiral building, after the Babel tower, is the *Minaret of Samarra's* mosque in Iraq, built in the VI century AD under the Abbasside dynasty and called Al- Malwiyya i.e. the spiral (see Fig. 1); it seems that Ibn Tulun Mosque (879 AD) in Cairo has been inspired by it .

The symbolic value of the spiral is evident and in the Muslim world it is directly related to the concept of infinity and hence of God; this is particularly true during the Abbasids Dynasty, indeed in that period, the traditional calligraphy is the Angular

Fig. 1 Minaret of Samarra's
mosque in Iraq, built in the
VI century AD under the
Abbasside dynasty and called
Al-Malwiyya

Fig. 2 Example of Angular
kufic writing (detail of a panel
from a Topkapi scroll)

kufic (see Fig. 2) which is based on the spiral and it was used as a decoration to cover
the Mosque and in the transposition of the koran. The idea being that the spiral gives
a centripetal movement of the reading, a rotational vision.

Always as an elevation toward God, but centuries later, it is impossible not to
mention Sant'Ivo alla Sapienza in Rome by the baroque architect Francesco Borromini. His mathematical and geometrical knowledge is well documented, and deeply
related to his deep religious belief. He repeatedly used geometrical concept to em-
body his quasi mystical faith and in particular Sant'Ivo's spiral suggests an elevation
towards infinity, in the research of Sapienza (knowledge) and God.

It is important to mention that this vision of the "spiral" is not only an occidental
vision since in the Hindu philosophy the powerful generating "Kundalini" force, is

Fig. 3 Left Sant'Ivo alla Sapienza, Francesco Borromini; right Tatlin's Monument to the Third International (1918)

by definition "coiled" in the spine and so is define through a variation of the term "Kundala" which means spiral in Sanskrit.

When the Soviet Socialist movement transplanted religious sentiments into ideology it was again the spiral that was used by architect Vladimir Tatlin in the construction of the Monument to the Third International (see Fig. 3). Soviet constructivism was used as a political propaganda for the construction of the new socialist society; Tatlin's tower was experimental, because though completely abstract, the double spiral was suppose to somehow represent the elevation of the russian people under the glorious prospective of the Bolshevik revolution.

Interestingly enough, Tatlin's tower inspired Roberto Semprini for its Tatlin's couch , the goal of the spiral is once more "reduced" as he says: *In today's mass consuming society, with the end of the great political utopias, this coach wants to realize a small utopia: let some art enter the living room[...] a revolution in the living room that in the boring design landscape, rises like a flag.*

2.2 Helicoids

The Helicoid is a two dimensional spiral whose parameter equation is

$$\begin{cases} x(s,\theta) = s\cos\theta \\ y(s,\theta) = s\sin\theta \\ z(s,\theta) = \theta. \end{cases}$$

It is far too obvious to mention that in architecture the geometrical surface denoted helicoid, which is somehow the natural extension of the spiral, has been used in infinite example of staircases. Precisely, the helicoid allows to solve a precise technical problem: rising while staying around a vertical axis. Of course it would be naive to

Fig. 4 Eero Saarinen staircase in the General Motors Technical Centre in Michigan (© Ezra Stoller / Esto)

think that its only value is "functional", in fact the complex and expressive geometrical structure has often played a central role in the formal, esthetic and symbolic characterization of the architecture involved.

We cannot help but go back to the mystical Borromini in its conception of the Barberini's staircase (he adds an elliptical plan to further emphasize the symbolic value of the staircase).

Eero Saarinen staircase in the *General Motors Technical Centre* in Michigan in its neo-expressionist design seems to use the "helicoid shape" in order to play with both light and engineering. It is antithetical to the used done by Herzog and De Meuron in the *Treppenhaus Bibliothek*, where the shape is used to enhance the expressionism through the use of colors "A building is a building. It cannot be read like a book; it doesn't have any credits, subtitles or labels like picture in a gallery. In that sense, we are absolutely anti-representational. The strength of our buildings is the immediate, visceral impact they have on a visitor" (Jaques Herzog, [4]).

On the other hand, in Frank Lloyd Wright's Guggenheim's museum the helicoid is the building itself and not only an element of it, hence revolutionizing both the role and the direction of the geometrical surface. Using Paolo Portoghesi's words *Borromini è stato forse l'architetto più innovatore degli ultimi secoli. Oggi è sentito come un precursore che ha prefigurato la condizione dell'architetto moderno. Molti hanno tratto ispirazione dalle sue opere. Come Frank Lloyd Wright, ad esempio: la spirale di S. Ivo si ritrova, rovesciata, nel museo Guggenheim.* [5] ("Borromini is maybe the most innovative architect of the last century. Today he is perceived as a pioneer that has pre-conceived the role of the modern architect. Many have been inspired by his opus. As Frank Lloyd Wright, for example: St. Ivo's spiral can be found upside down in the Guggenheim museum").

Fig. 5 An hyperboloid of one
sheet, see the straight lines

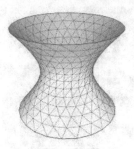

3 Ruled surfaces

A ruled surface is a surface such that through each of its points passes a straight line
contained in the surface, so that it can parametrized by

$$\begin{cases} x(t,s) = a_1(s)t + b_1(s), \\ y(t,s) = a_2(s)t + b_2(s), \\ z(t,s) = a_3(s)t + b_3(s). \end{cases}$$

For obvious structural reasons, ruled surfaces are convenient geometrical objects
to use in architecture and so have been used throughout the centuries. One such
example is the Helicoid which was treated above which is just a straight line that
turns around a pole. Interestingly, to design or build a ruled surface it is often enough
to construct a curve and then an infinite number of the straight lines (beams) through
that curve. And whether it is reinforced concrete or steel beams, the structural primal
elements are linear; hence ruled surfaces are extremely convenient shapes to use in
term of time and price. This applies also to steel and glass.

3.1 Hyperboloids

Of particular interest is the one sheet hyperboloid (see Fig. 5) which is just a set of
straight lines around a circle and perpendicular to it that are twisted, its equation is
given by $x^2 + y^2 - z^2 = 1$. In fact it is a ruled surface such that at each point there
are two straight lines contained in the surface.

The hyperbolic hyperboloid is also a ruled surface. Antonio Gaudì, designing
the *Sagrada Familia*, used many ruled surfaces exactly for the reasons mentioned
above, in his gothic and organic vision of architecture, he used geometrical structure
present in nature, not for imitation but in a careful planning of the convenient use
of the generating forces that are behind some "natural" constructions. Hence many
pillars of the Sagrada Familia are hyperbolic hyperboloids that are in fact inspired
by trunk trees [9].

Fig. 6 Oceanografic by Felix Candela, Valencia

Of course the list of hyperbolic hyperboloids in architecture and design is very long we shall just mention the *Cathedral* in Brasilia by Oscar Niemeyer, *McDonnell Planetarium* in Saint Louis by Gyo Obata, *Coop Himmelblau's BMW* building, or the *Falkland lamp* by Bruno Munari (though the latter is most likely a catenoid.).

3.2 Other ruled surfaces

It would be reductive to think that the helicoids and the hyperbolic hyperboloid are the only ruled surfaces present in architecture, there are others even more complex like: *Le Corbusier's* "Phillips pavilion" in Brussel, Felix Candela's Oceanografic in Valencia (see Fig. 6) , Toyo Ito's *Relaxation Park* in Spain, James Stirling's *New State Gallery* in Stuttgart, Nicoletti's new *Hall of Justice* in Arezzo etc.

4 Fiber bundles

In most examples of ruled surfaces, one can describe the surface as a straight lines that moves around another curve, but of course there are no obligations to consider only straight lines, one can take any family of curves that fibrate on a curve, a simple example is the tori, which is just a fibration of circles perpendicular to a given circle.

In general it is possible to fibrate surfaces on surfaces, the most important fibration is the tangent fiber bundle, which is the set of planes that are tangent to a surface, because it allows to do differences and hence "calculus" on curved surfaces. The visualization of fiber bundle is obviously not easy, but it is particularly understandable seeing Roberto *Capucci's Bolero* [3] or Norman Foster's dome on top of the Berlin's *Reichstag* building), but in the *Artichoke Lamp* by Poul Henningsen (see Fig. 9) the planes that fibrates the surface are not tangent to it.

Fig. 7 Poul Henningsen, *Artichoke Lamp*

5 Genus of a surface or architecture with holes

The tori, beside being a fibration is an example of surface of genus 1 i.e. with one hole, mathematicians can give a precise notion of hole, and through it, thanks to the theorem of Gauss Bonnet, give a precise relationship between the genus of a closed surface M and its area weighted by its curvature:

$$\int_M Kds = 2\pi\chi(M)$$

where K is the curvature and $\chi(M)$ is its Euler characteristic i.e. $\chi(M) = 2 - 2g$ where g counts the number of holes.

Two-topology is the science that allows to classify the surfaces by counting their holes. Recent architectural examples of surfaces with "holes" are: in Milan *Trade Fair* by Fuksas, *Burnham Pavilions* in Chicago (one by UN Studio and one by Zaha Hadid), Toyo Ito's Grin Grin Park in Fukuoka or his project for the Forum for Music Dance and Visual Culture in Ghent Belgium (designed together with Andrea Branzi).

6 Inspirations

We recall that the curvature of a surface measures the mean of the highest and the smallest of the curvatures of the curves obtained by intersecting of the surface with an orthogonal plane; e.g. the curvature of the sphere is just the inverse of the square of the radius while the curvature of a plane or a cylinder is null everywhere. Surfaces

Fig. 8 Dini's surface

with negative curvature are somehow more disturbing then those with positive curvature, since the negative curvature suggest a saddle, so surfaces where each point can be seen as a saddle are rare and create an aesthetic tension, one such example is the Dini's surface another one is the hyperbolic hyperboloid. It is interesting to wonder if Hadid's *Nordpark Cable Railway* in Innsbruk is a piece of Dini's surface or just resembles it i.e. which is the awareness of Hadid in the use of it.

Another interesting example is Musmeci's bridge in Potenza (see Fig. 9), where the negative curvature is not only used in order to enhance the aesthetic power of the pillars but to solve an engineering problem i.e. to englobe in a continuous surface both the horizontal and the vertical structural elements.

An algebraic surface is a set of points whose coordinates annihilate a polynome. The possible shapes taken by algebraic surfaces are of infinite kind and complexity, here we shall show a few that have particularities that could be of inspiration for architectural applications.

Fig. 9 Bridge on the Basenta, by Ing. S. Musmeci. Courtesy of P. Musmeci

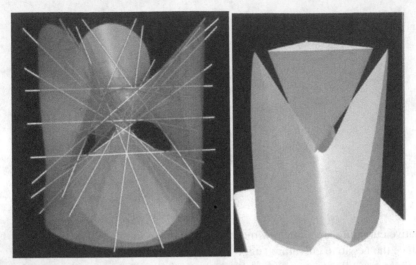

Fig. 10 Diagonal Cubic Clebsch, with its Salomn seal lines and S3 by Gianmarco Todesco

We start by Todesco's favorite algebraic surface , the *Diagonal Cubic Clebsch* (see Fig. 10) whose equation is

$$16x^3 + 16y^3 - 31z^3 + 24x^2y - 48x^2z - 48xy^2 + 24xz^2 + 51\sqrt{3}z^2 - 72z = 0.$$

It is characterized by the fact that it contains 27 straight lines (Salomon's seal lines), implying that even though it is not a ruled surface it has some characteristics of ruled surfaces that could make it constructible through beams.

Another cubic is known as S3 its equation is

$$9x^3 + 3x^2z - 4x^3 - 27y^2x + 3y^2z - 4/7(9x^2 + 9y^2 - 4z^2)(2z + 1) = 0.$$

Using Todesco's description it is obtained through a homographic transformation of Cayley's surface, the singular points create an interesting anomaly.

Minimal surfaces are surfaces with constant mean curvature, one example is given by Schwarz PD surface, which is was constructed and named by A. Schoen (1970) [6] as an example of triply periodic minimal surface.

Fig. 11 Triply periodic Minimal Surface

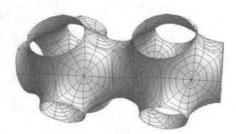

References

1. A. Capanna, Conoids and Hyperbolic Paraboloids in Le Corbusier's Philips Pavilion, pp. 35-44 in Nexus III: Architecture and Mathematics, ed. Kim Williams. Pisa, Pacini Editore, 2000. http://www.nexusjournal.com/conferences/N2000-Capanna.html
2. N. Manfredi, Sergio Musmeci. Organicità di forme e forze nello spazio. Testo e Immagine, Torino, (1999).
3. I. Birindelli, Superfici di Seta: la geometria negli abiti di Capucci. In: M. Emmer (ed.) Matematica e cultura 2010. Springer, Milano, 2010, 67-70.
4. An interview, quoted in http://www.greatbuildings.com/architects/Herzog_and_de_Meuron.html.
5. P. Portoghesi. Francesco Borromini. Mondadori Electa, Milano, 1989.
6. A. Schoen, Infinite periodic minimal surfaces without self-intersection. NASA Technical Note D-5541, 1970.
7. R. Semprini, E tutto cominciò dalla spirale. I quaderni di Lang, Silver books, 2001.
8. G. Todesco, Le forme della matematica, http://www.dm.unito.it/modelli/
9. M. Burry, Gaudì l'innovatore. In: M. Emmer (ed.), Matematica e cultura 2004. Springer, Milano, 2004, 143-170.

References

Visual Harmonies:
an Exhibition on Art and Math

Michele Emmer

1 Art and Math: the idea of space

Art and Mathematics are inextricably linked. One of the main linkages is the idea
of space and its transformations. There is no doubt that throughout the history of
Western culture, there were moments when the relationship between mathematics
and art has remained hidden, if not absent, and others in which it appeared with
great vividness. And one of the privileged periods is obviously the *Renaissance*.

> For several reasons the problem of depicting the real world led the Renaissance painters to
> mathematics. The first reason was one that could be operative in any age in which the artists
> seek to paint realistically. Stripped of colour and substance the objects that painters put on
> canvas are geometrical bodies located in space [...].

Fig. 1 MART (Museo Arte
Moderna Rovereto), archi-
tect Mario Botta, 2002

Michele Emmer
Department of Mathematics, Sapienza University of Rome (Italy).

Emmer M. (Ed.): Imagine Math. Between Culture and Mathematics
DOI 10.1007/978-88-470-2427-4_11, © Springer-Verlag Italia 2012

The Renaissance artist turned to mathematics not only because he sought to reproduce nature but also because he was influenced by the revived philosophy of the Greeks. He became thoroughly familiar and imbued with the doctrine that mathematics is the essence of the real world, that the universe is ordered and explicable rationally in terms of geometry [1].

So wrote the math historian Morris Kline in the book *Mathematics in Western Culture* and he added that even less well known is the fact that:

... mathematics has determined the direction and content of much philosophic thought, has destroyed and rebuilt religious doctrines, has supplied substance to economic and political theories, has fashioned major painting, musical, architectural and literary styles, has fathered our logic [...]. Finally, as an incomparably fine human achievement mathematics offers satisfactions and aesthetic values at least equal to those offered by any other branch of our culture.

The great master of Perspective and one of the best mathematicians of the *Quattrocento* was Piero della Francesca. His text *De perspectiva pingendi* was written in 1474.

Martin Kemp writes that to understand the work of Piero as a scholar of perspective, we should remember that he had a deep knowledge of pure and applied mathematics, enough to write treatises of sufficient quality to compete with any work in the Italy of his time.

Years after Piero, precisely in 1623, Galilei wrote in *the Essayer* (Il Saggiatore) [2]:

It seems to detect a firm belief that, in philosophising, it is necessary to depend on the opinions of some famous author, as if our minds should remain completely sterile and barren, when not wedded to the reasoning of someone else. In philosophizing one must support oneself upon the opinion of some celebrated author, as if our minds ought to remain completely sterile and barren unless wedded to the reasoning of some other person. Possibly he thinks that philosophy is a book of fiction by some writer, like the Iliad or Orlando Furioso, productions in which the least important thing is whether what is written there is true. Well that is not how matters stand. Philosophy is written in this grand book, the universe, which stands continually open to our gaze. But the book cannot be understood unless one first learns to comprehend the language and read the letters in which it is composed. It is written in the language of mathematics, and its characters are triangles, circles, and other geometric figures without which it is humanly impossible to understand a single word of it; without these, one wanders about in a dark labyrinth.

Thus, without mathematical structures we cannot understand nature. Mathematics is the language of nature.

A few centuries, in 1904 a famous painter wrote to Emile Bernard [3]:

Traiter la nature par le cylindre, la sphère, le cône, le tous mis en perspective, soit que chaque cote d'un objet, d'un plan, se dirige vers un point central. Les lignes parallèles à l'horizon donnent l'étendue, soit une section de la nature. Les lignes perpendiculaires à cet horizon donnent la profondeur. Or, la nature, pour nous hommes, est plus en profondeur qu'en surface, d'oμu la nécessite d'introduire dans nos vibrations de lumière, représentée par les rouges et le jaunes, une somme de bluetes, pour faire sentir l'air.

The art historian Lionello Venturi commented that in Cezanne's (the artist in question) paintings there are no cylinders, spheres and cones, so the artist's quote represents nothing but an ideal aspiration to an organization of shapes transcending nature.

During the period when Cezanne was painting, and even a few years earlier, the panorama of geometry had changed since Galileo's time. In the second half of the 19th century geometry had mutated significantly. In a letter of December 1799 Gauss wrote to Farkas Bolyai on his tentative to prove the Fifth Postulate of the Elements of Euclid, starting from a demonstration by absurd: "My works are very advanced but the way in which I am moving is not conducing to the aim I am looking for, and that you say to have reached. It rather seems to put in doubt the exactness of geometry." Gauss never published his results on this particular topic in his lifetime. In 1827 he published the "Disquisitiones generales circa superficies curves" in which he introduce the idea of studying the geometry of a surface in a "local way" without minding its immersion in a three dimensional space, studying the invariant properties of the surfaces. He also introduces the idea of the curvature of the surface. Between 1830 and 1850 Lobacevskij and Bolyai built the first examples of non-Euclidean geometry, in which the famous fifth postulate by Euclid was not valid. Not without doubt and conflicts, Lobacevskij would later call his geometry (which today is called non-Euclidean hyperbolic geometry) *imaginary* geometry, because it was in such strong contrast with common sense. For some years non-Euclidean geometry remained marginal to the field, a sort of unusual and curious form, until it was incorporated into and became an integral part of mathematics through the general ideas of G.F.B. Riemann (1826-1866). In 1854 Riemann held his famous dissertation entitled *Ueber die Hypothesen welche der Geometrie zur Grunde liegen* (On the hypotheses which lie at the foundation of geometry) before the faculty of the University of Göttingen (it was not published until 1867). In his presentation Riemann held a global vision of geometry as the study of varieties of any dimension in any kind of space. According to Riemann, geometry didn't necessarily need to deal with points or space in the traditional sense, but with sets of ordered n-ples.

In 1872 in his inauguration speech after becoming professor at Erlangen (known as the *Erlangen Program*), Felix Klein (1849-1925) described geometry as the study of the properties of figures with invariant character in respect to a particular group of transformations. Consequently each classification of the groups of transformations became a codification of the different types of geometry. For example, Euclidean plane geometry is the study of the properties of the figures that remain invariant in respect to the group of rigid transformations of the plane, which is formed by translations and rotations.

Jules Henri Poincaré held that:

... the geometrical axioms are neither synthetic a priori intuitions nor experimental facts. They are conventions. Our choice among all possible conventions is guided by experimental facts; but it remains free, and is only limited by the necessity of avoiding every contradiction, and thus it is that postulates may remain rigorously true even when the experimental laws, which have determined their adoption, are only approximate. In other words the axioms of geometry are only definitions in disguise. What then are we to think of the question: Is Euclidean geometry true? It has no meaning. We might as well ask if the metric system is true and if the old weights and measures are false; if Cartesian coordinates are true and polar coordinates are false. One geometry cannot be more true than another; it can only be more convenient. Euclidean geometry is and will remain the most convenient.

Poincaré, in *Analysis Situs* (Latin translation of the Greek), published in 1895, is also responsible for the official birth of the sector of mathematics that today is called *Topology*: "As far as I am concerned, all of the various research that I have performed has brought me to *Analysis Situs* (literally analysis of place)". Poincaré defined topology as the science that introduces us to the qualitative properties of geometric figures not only in ordinary space, but also in more than 3-dimensional space. Without any doubt, one of the main contributions of mathematics to art is the transformation, the mutation of the idea of space.

Years later an artist of the twentieth century clarified [4]:

> By a mathematical approach to art it is hardly necessary to say I do not mean any fanciful ideas for turning out art by some ingenious system of ready reckoning with the aid of mathematical formulas. [...] It must not be supposed that an art based on the principles of mathematics is in any sense the same thing as a lastic or pictorial interprepation of the latter. Indeed it employs virtually none of the resources implicit in the term pure mathematics. The art in question can best be defined as the building up of significant patterns from the everchanging relations, rhythms and proportions of abstract forms, each one of which, having its own causality, is tantamount to a law unto itself. As such, it presents some analogy to mathematics itself. Just as mathemaics provides us with a primary method of cognition, and can therefore enable us to apprehend our physical surroundings, so, too, some of its basic elements will furnish us with laws to appraise the interactions of separate objects, or groups of objects, one to another and again, since it is mathematics which lends significance to these relationships, it is only a natural step from having percieved them to desiring to portray them.

So wrote Max Bill in 1949 to trace the lines of a possible mathematical approach to modern art. Bill was one of the artists partecipating in an unusual exhibition. From January 20 to February 17, 1963 an art exhibition was held in Paris. It was unusual, first for its location: the *Palais de la Découverte*, the temple of the popularization of science in France until the opening in the early eighties of the *Cité des Sciences de la Villette*. An art exhibition entitled *Formes mathématiques, Painters, Sculptors contemporains.*

Works of artists of great importance were exhibited, among which among Max Bill, Paul Cezanne, Robert and Sonya Delaunay, Albert Gleizes, Juan Gris, Le Corbusier, Jean Metzinger, Piet Mondrian, Laszlo Moholy-Nagy, Georges Seurat, Gino Severini, Sophie Tauber-Arp, Victor Vasarely. Among the sculptors Max Bill, Raymond Duchamp-Villon, Georges Vantongerloo. The exhibition was organized in three sections: *Mathématiques, Peintres, Sculptors*. In the first section numerous mathematical surfaces made of metal or plaster were exposed.

The middle pages of the small catalog had four illustrations: two geometric surfaces, and works by Robert Delaunay, Barbara Hepworth and Gino Severini. The introduction to the exhibition was by Paul Montel, mathematician, curator at that time of the mathematics section of the *Palais de la Découverte*. Montel wrote [5]:

> It may seem surprising that there are relationships between art and mathematics, between the world of quality and quantity of the world. Nevertheless, close connections link together these two different worlds of representation. [...] In fact, each of these two activities, mathematical research and artistic creation, is in debt to the other. An important result obtained in mathematics, to its author offers an aesthetic joy similar to that which can give the architectural harmony or musical chords.

Interactions, proportions, transformations and harmony, as well matemorphosis and their metaphoric meaning became central concepts in the relation between art, math and architecture too. It is not by chance that the theme of the Biennale of Architecture in Venice in 2004 was *Metamorph* [6]:

> Many of the great creative acts in art and science can be seen as fundamentallymetamorphic, in the sense that they involve the conceptual re-shaping of orderingprinciples from one realm of human activity to another visual analogy. Seeing something as essentially similar to something else has served as a key tool in the fluid transformation of mental frameworks in every field of human endeavour. I used the expression 'structural intuitions' to try to capture what I felt about the way in which such conceptual metamorphoses operate in the visual arts and the sciences. Is there anything that creators of artefacts and scientists share in their impulses, in their curiosity, in their desire to make communicative and functional images of what they see and strive to understand? The expression 'structural intuitions' attempts to capture what I tried to say in one phrase, namely that sculptors, architects, engineers, designers and scientists often share a deep involvement with the beguiling structures apparent in the configurations and processes of nature - both complex and simple. I think we gain a deep satisfaction from the perception of order within apparent chaos, a satisfaction that depends on the way that our brains have evolved mechanisms for the intuitive extraction of the underlying patterns, static and dynamic.

These are the words of Martin Kemp, an art historian specialized in the relationship between art and science in the article - *Intuizioni strutturali e pensiero metamorfico nell'arte, architettura e scienze*, in *Focus*, one of the volumes that made up the catalogue of the 2004 Venice International Biennale of Architecture. In his article Kemp writes mainly about architecture. The image accompanying Kemp's article is a project by Frank O. Gehry, an architect who obviously cannot be overlooked when discussing modern architecture, continuous transformation, unfinished architecture, and infinite architecture.

Fig. 2 The logo of MART

2 An exhibition on Art and Math: the idea of space

Adding to all this the geometry of complex systems, fractal geometry, chaos theory and all of the "mathematical" images discovered (or invented) by mathematicians in the last thirty years using computer graphics, it is easy to see how mathematics has contributed to changing our concept of space - the space in which we live and the idea of space itself. Because mathematics is not merely a means of measurement in recipes, but has contributed, if not determined, the way in which we understand space on earth and in the universe.

These are some of the reasons behind a major exhibition dedicated to the idea of space and the relationships between modern and contemporary art and mathematics from the twentieth century to the present day at the MART in Rovereto, one of the important European Musems of Modern and Comteporary Art. The exhibition *Visible harmonies: the idea of space between art and mathematics*, will probably open March 2013 and will remain open until the end of June 2013 [7].

The general idea of the exhibition is by Michele Emmer, the project by the Director of the MART, and Michele Emmer. The other members of the scientific committee are: Umberto Bottazzini, Università di Milano, Linda D. Henderson, University of Texas at Austin, Michael Rottman, Frei Universität Berlin. Curators of the catalogue, in two different volumes, italian and english: the Director and Michele Emmer.

References

1. M. Kline, Mathematics in Western Culture. Penguin Books, Harmondsworth, 1953, p. 150.
2. G. Galilei, Il saggiatore, Rome, 1623; English translation: The Assayer, by Stillman Drake. In: Discoveries and Opinions of Galileo. Doubleday & Co., New York, 1957, pp. 231-280.
3. L. Venturi, La via dell'impressionismo: da Manet a Cezanne. Einaudi, Torino, 1970, pp. 268-269.
4. M. Bill, A Mathematical Approach to Art, 1949; reprinted with the author's corrections. In: Michele Emmer (ed.), The Visual Mind: Art and Mathematics. MIT Press, Boston, 1993, pp 5-10.
5. P. Montel, L'Art et les Mathématiques. In: J. Cassou, P. Montel (eds.), FORMES mathèmatiques, peintres, sculpteurs contemporaines. Universitè de Paris, Palais de la Decouverte, Paris, 20 Janvier – 17 février 1963. Courtesy of Marjorie Malina.
6. M. Kemp, Intuizioni strutturali e pensiero metamorfico nell'arte, architettura e scienze. In: K.W. Forster (ed.), Metamorph: Focus, catalogue. La Biennale d'Arte di Venezia. Marsilio, Venezia, 2004, pp. 31-43.
7. http://www.mart.it
   M. Emmer, Visibili armonie: arte, cinema, teatro e matematica. Bollati Boringhieri, Torino, 2007.

Emilio Prini, Alison Knowles, and Art's Logic

Cornelia Lauf

This is a short description of how art works. For those who know math (I don't), go figure.

In the period between 1968 and 1974, artist Emilio Prini executed several hundred drawings on plain white paper, fabricated with the help of an Olivetti 22 typewriter. Some look like architectural drawings, others like mathematical formulas, poems, or musical scores. The typewriter was twisted every which way for its repertoire of possible visual images. Prini forced the machine into making a "1001 nights" of drawings.

These studies (more than three hundred) are currently being worked up into a book for Three Star Books, a publishing house founded by myself and two colleagues, Christophe Boutin and Mélanie Scarciglia, in Paris. The volume is entitled *Proporzione 2/7*.

Emilio Prini is one of the elder statesmen of Italian postwar art. Voracious reader, connoisseur of music, literature, art, architecture, and food, among many other disciplines. Diligent in the elaboration of one of the least understood and most elusive of postwar artistic practices.

Although it seems self-congratulatory to discuss one's own publication with an artist, and to thus both produce and publish one's own ideas oneself, this is the essence of making artworks. You have to believe in the whole thing, hook, line, and sinker. Thus, the description of two artworks – two books in this case – shall stand here in the context of this volume of essays, in the place of a more abstract discussion on the growth/creation of an artwork.

There is no real rationale for my doing this, and indeed much of what one does in life, as in art, seems to run counter to logic. And yet, there is urgency, especially on the part of great artists, to proceed, even if it appears as folly, simply because "geniuses create because they must."

There are indeed formulas, codes, order and reasons for creation, and they are known well by those inside the center of art. This is a topic of debate, particularly as it appears that increasingly it is possible to map codes – even mathematically or electronically – to identify elements as seemingly subjective as literary style. The logic inside chaos is known to those who pay with their skin for their decisions

Cornelia Lauf
IUAV University, Venice (Italy).

Emmer M. (Ed.): Imagine Math. Between Culture and Mathematics
DOI 10.1007/978-88-470-2427-4_12, © Springer-Verlag Italia 2012

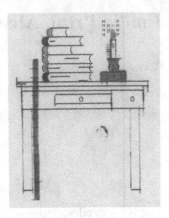

Fig. 1 Emilio Prini, drawing,
ink on paper, 1974 (detail)

to live art in their every waking moment. It means caring about philosophy and meaning so deeply that the only way to deal with it is "aesthetically" (as many might say) or materially, for sure.

Emilio Prini's collection of drawings is entitled *Proporzione 2/7* because it is a selection of proportional studies, many of which deal with the relation between the numbers 2 and 7. There are a few earlier drawings. Many are dated. These drawings can be arranged in chronological order. There is a section of so-called "scarti," rejected drawings.

The more than three hundred drawings are divided into three sections, in which numerous themes are worked out, even "checked off," when a certain theme has been explored in depth and resolved conceptually. Sometimes, this resolution takes place over a sequence of pages, as in the case of the drawings resembling building blocks, like *Vecchia fortezza* (the old fortress), or in the case of *La Scala* (the ladder). These pages function almost like a flip-book. An antecedent to this use of "motion" picture editing may be found in the 1970 *Arte Povera* catalogue of Germano Celant, in which photographs of the artist progress from a picture of his half-naked body to a photo of the artist's measure of his own body weight in the form of cubes made out of lead. This magnificent sequence, extraordinary in its modesty, and density, measures the ineffable moment that artists from Xeuxis to Bernini dealt with, of the transformation of stone into life (or vice versa). Vying neck to neck with the Fibonacci research of Mario Merz, the ability to make a bicycle seat into a bull, or ceramic urinal into a sculpture, Mr. Prini has gone one up, and in his utter anonymity, taken on the formulas underlying life itself.

In a first photo in the Celant book, there is a black-and-white image of the young artist, waking in his bed, and rising out of a plaid blanket. The image was shot by a Genovese friend of Prini's. This man went to Prini's house at dawn, and photographed him right as he awoke. The artist's alabaster skin, rising out of the white sheets, and his aquiline profile and mutton chop sideburns, resemble a sculpture by Canova, or a painting by Ingres.

It's just a photograph, though, a medium that Prini has explored so relentlessly as to include even its destruction, through over-usage, in one work/performance, until the camera gave up the ghost.

In other artworks, Prini is less visceral. He has created images about the viewing act and virtuality, in which a camera is depicted frontally, as the subject itself, in a mirror effect. Rather than being just the means toward making a reproduction, the camera is pictured as the subject. Forty years before the current replacement of text through image culture, Prini set up tautological situations regarding self-presentation and mediatization.

Another early work depicts a television, full screen, black and white, again the subject of the piece. In a way, these machines are self-portraits. The artist as machine.

The body of Prini is like a machine too, pushed to function and produce. Although the artist is famous for saying *Non faccio mai niente*, and *Non sono un artista* he is a relentless worker and thinker, subjecting his mind and body to the utmost fatigue: a life dedicated totally and utterly to art.

At the heart of Prini's projects are a desire to find the mechanisms that underlie biological process - a concern that typifies the work of many of the founders of *arte povera*.

And strangely enough, it is this play with the threshold of life, and making even lived life be part of artistic investigation - turning the eating of a piece of organic bread with an anchovy on top into an artistic experience - that the moment of making art lies.

While I was working on the publication with Emilio Prini in Rome, in which dozens of drawings created with the Olivetti typewriter chronicle the passage of written script into organic forms, I began a project with another artist, Alison Knowles, in New York. Knowles is one of the founders of the movement, Fluxus, and a wonderfully gifted performer, whose tributes to everyday process and the simple beauty of life shapes - from the shape of beans to the making of a sandwich - has been a feature of her work for over fifty years.

Oddly enough, Knowles' project too was about the process of creation. First, Knowles proposed to make a book using a series of phrases subsequently scrambled by computer and turned into a long and mesmerizing poem. This method and the concept of Knowles' initial work was based on her computer-generated book/computer score *House of Dust* published by Verlag Gebr. König, Cologne, in 1969: the first literary score created by a machine.

By sheer coincidence, on the other side of the world, Knowles' project, produced by computer rather than the typewriter that Prini used, involves the Fibonacci cycle too. Knowles conceived of phrases in three disparate categories: "situations," "weather/time," and "place." These were then cross-matched (like a *Cadavre Exquis* game) into all possible permutations. A 64-page book published by one star press ensued. However, the actual design of the project (now that computers no longer necessarily spit out long rolls of green perforated paper, like in 1969) was up for grabs. So, I asked artist Rirkrit Tiravanija (sympathetic to Fluxus, afficionado of the work of Knowles) to take over this element. The first suggestion he made was

MEN AND WOMEN COMMONLY DRESS ALIKE

WHEN STORMS ARE SEEN

COMING IN UNDERGROUND.

MEN AND WOMEN COMMONLY DRESS ALIKE

WHEN STORMS ARE SEEN

COMING IN OVER THE HILLS AND THROUGH THE WOODS.

MEN AND WOMEN COMMONLY DRESS ALIKE

WITH CLEAR SKIES

ALL WEEK UNDERGROUND.

MEN AND WOMEN COMMONLY DRESS ALIKE

WHEN TORRENTIAL RAIN

IS EXPECTED BY NIGHTFALL UNDERGROUND.

MEN AND WOMEN COMMONLY DRESS ALIKE

WHEN SNOW BLIZZARDS ARE PREDICTED

IN A PLACE WITH ROCKS AND SAND, GOOD FOR WALKING.

SMALL BIRDS AND FLOWERS ARE UBIQUITOUS

WITH CLEAR SKIES ALL WEEK

IN A PLACE WHERE EVERYONE PLAYS OUTDOORS.

WEALTH IS EXPOSED FOR ITS FAMILY CONNECTIONS

WITH CLEAR SKIES ALL WEEK

ON A PENINSULA WITH THE OCEAN ON ONE SIDE AND BAY ON THE OTHER.

WORK IS NOT CENTRAL TO DAILY ACTIVITY

WHEN TORRENTIAL RAIN IS EXPECTED BY NIGHTFALL

OVER THE HILLS AND THROUGH THE WOODS.

PIGS ARE ALLOWED TO ROAM FREE

WHEN SNOW BLIZZARDS ARE PREDICTED ON A PENINSULA

WITH THE OCEAN ON ONE SIDE AND BAY ON THE OTHER.

THE UNDERWORLD AND THE OVERWORLD NEVER TOUCH

WHEN TORRENTIAL RAIN IS EXPECTED BY NIGHTFALL

ON A PENINSULA WITH THE OCEAN ON ONE SIDE AND BAY ON THE OTHER.

OLD MEN GATHER IN GROUPS TO DISCUSS THE WEATHER

WITH CLEAR SKIES ALL WEEK

IN A PLACE WITH ROCKS AND SAND, GOOD FOR WALKING.

REAL INFORMATION IS EVERYWHERE AVAILABLE

BUT OF LITTLE INTEREST

WITH FIERCE WIND AND BITTER COLD IN FRANCE.

Fig. 2 Alison Knowles and Rirkrit Tiravanija, "Men and Women Commonly Dress Alike" (detail) (Paris, Three Star Books, 2011)

to set Knowles' text in his characteristic typeface, and the second was to scramble Knowles' poem again, this time adding the filter of the Fibonacci system (1, 2, 3, 5, 8, 13, and so forth). The oddly short result has been presented in a tribute to the Eastern book, in the form of a digitally printed scroll with two bamboo rods, devised using canvas selected by Knowles.

Neither the Prini nor the Knowles/Tiravanija project occurred with knowledge of the other, yet the artists were all working with formulas that might describe natural process.

Entropy and chance, but also some strange form of symmetry and determination, are at the heart of art. This small text demonstrates the uncanniness of the process, and the randomness as well as intent that go into making artworks–two books in this case–with artists whose work capture the rhythm of life processes.

All the Numbers End in Numbers.
On a Work by Alighiero Boetti

Andrea Valle

Alighiero Boetti is one of the most representative contemporary Italian artist and his opus is raising a constantly growing international interest. Many of his works can be realized on very different supports and make use of algorithmic procedures. This contribution analyzes a minimal work, consisting only of the linguistic description of a process, with the aim of demonstrating that, in spite of its simplicity, it shows various relevant features of Boetti's aesthetics.

1 Sensing numbers: Alighiero Boetti

Alighiero Boetti (1940-1996) is one of the most prominent artistic figures in the cultural landscape of conceptual art. After a first working period [1] that can be traced back to the "Arte povera" movement [2], he started a highly personal path, which has lead him to become more and more influencing during recent years, raising a growing interest in his work and poetics. While Conceptualism tends to strongly remark the role of the idea against the material qualities of the resulting work, of the project against its implementation, the situation is more subtly complex in Boetti, as abstract conceptual devices are intended to generate sensible outputs without favoring an aspect against the other. As an example, *Alternando da 1 a 100 e viceversa* ("alternating from 1 to 100 and vice versa", hence on *Alternando*) is the name of a series of works by Alighiero Boetti that represents at its best some features of Boetti's art. Despite its absolutely simple structure, or maybe exactly because of this aspect, it is a particularly elegant and clear statement of Boetti's aesthetics[1]. In *Alternando* (see Fig. 1) the starting point is a chessboard consisting of 100 squares; each square consists of other 100 small squares. The chessboard is the theatre of a numerical progression; in the first square one of the small squares is left white, while all the others are black; in the second square there are two black small squares and 98 white; in the third square 97 small squares are black and 3 white; ... in the penultimate square

---

Andrea Valle
CIRMA, University of Torino (Italy).

[1] For a general presentation see [3], for a formal and semiotic description [4] and [5].

Emmer M. (Ed.): Imagine Math. Between Culture and Mathematics
DOI 10.1007/978-88-470-2427-4_13, © Springer-Verlag Italia 2012

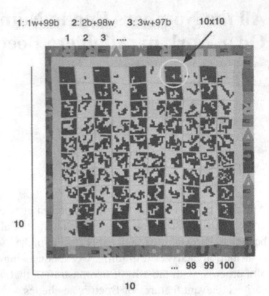

Fig. 1 Filling mechanism in Boetti's *Alternando da 1 a 100 e viceversa*, embroidery on canvas, 1977 ("w": white "b": black)

there are 99 white small squares; the last square is completely blackened ([1], 154, for a description by Boetti emphasizing the combinatorial pathos see [1], 204).

The decision about which small squares in each square are to be blackened or to remain white is left undetermined by the definition, and it has to be taken by the implementing subject (that, as typical with Boetti, may not coincide with the artist himself). On one side, the formal specification of the work simply assumes two graphic primitives: square drawing and binary colorization (black/white). This abstractness is typical of Boetti's love for hand drawing (often with pencil or pen) as a way of visualizing patterns and processes[2]. But this is just a part of the Boettian practice. As it can be seen on Fig. 1, the filling mechanism operates sequentially on the chessboard visually exploiting the two-state inversion mechanism. The abstractness of the design is in fact intended to allow for multiple implementations on very different supports. Some versions of the work are hand-drawn with pencil by Boetti himself, but other are embroidered (as the version presented in Fig. 1), weft on kilims, realized as wall mosaics or as a large square pavement. In Fig. 1 it can be noted that the colored border, even if very often present in mostly *Alternando*'s realizations (as a sort of counterpart to the black/white progression), is not required for the attribution of a work to the series. All these cases witness the relevance that Boetti assigns to the final output, in order to avoid the perils of an idealistic drift which is somewhat typical of Conceptualism. This "sensible Conceptualism" –so to say– remarks both the role of the mental process at the source of the art work and the relevance of its concrete realization through a material practice. By coupling

---

[2] See *Storia naturale della moltiplicazione*, *Cimento dell'armonia e dell'invenzione*, *Autodisporsi* for pencil, or the so-called *Lavori a biro* (biro works), realized with pen [1], [6].

these two poles, the world itself reveals an unsuspected richness to the one able to see it, not in terms of a mysterious quality to be recognized by the adept of a mystic depth, but in terms of a capability of reading the surface of the phenomena, with the intermingling play of order and disorder, to use an opposition which gave the title to a large series of works by Boetti (*Ordine e disordine*, see [1], 136-9 and [6], 147-9). It is worthwhile to investigate all the previously introduced aspects by discussing a real minimalist work, which focuses on the conceptual and procedural side of Boetti's poetics but that will reveal at the end its intimate relationship with a phenomenology of the world.

2 All the numbers end in three

Boetti can candidly assume that "the number is the only real entity existing in the universe" ([6], 205). But it must be noted that in the artist's opus this Platonism seems to turn into Pythagoreanism, as the number is not intended by Boetti as a static entity but exactly as "a form unfolding in the becoming" ([7], 22). Boetti's fascination for numbers refuses any compromise with numerology, favoring instead the pure exhibition of a numerical progression, algorithmically defined. His genuine amazement for numbers is indeed evident in the *Alternando*'s construction, where it seems to be possible to literally hear the Pythagorean definition of number reported by Stobaeus as "a progression of quantity beginning from the one and a regression that ends in it" ([7], 22). But this idea of the number as a form that unfolds in the becoming probably finds its purest example in a crystal-clear "little-known work" *Tutti i numeri finiscono in tre* [All numbers end in three]. Giovan Battista Salerno, one of the most acute commentators of Boetti's work and poetics, explains in Italian (as we will see, the language is relevant):

> Si prenda per esempio il numero otto: è fatto di quatto lettere; quattro è di sette lettere; sette è di cinque lettere; cinque è di sei lettere; sei è di tre lettere e il tre conta le proprie stesse lettere. Dunque otto finisce in tre. Questo sistema funziona con *tutti* i numeri, per grandi che siano. Si tratta solo di alternare i numeri e le lettere, leggere e contare. Si noti che quest'opera, un'opera piccolissima che con il suo modo leggero e giocoso aggiunge forse qualcosa al grande sistema simbolico del numero tre, un'opera come questa *non si vale di alcun supporto materiale* [8][3].

This work, which has no place in Boetti's catalog, is genuinely Boettian and coincides with its linguistic description (provided by Salerno after one of the usual conversations with the artist). Its clear algorithmic nature resembles *Alternando*, but while the latter masterpiece is embodied by a whole system of implementations, that

---

[3] Take for example the number otto [eight]: it is composed of four letters, four [quattro] is composed of seven letters, seven [sette] of five letters, five [cinque] counts six letters, six [sei] counts three letters and three [tre] counts its own letters. Thus, eight ends in three. This system works with *all* the numbers, no matter how large. It is enough to alternate numbers and letters, to read and to count. Note that this work, a work which, with its light and playful mood, perhaps adds something to the great symbolic system of the number three, a work like this *does not use any material support*.

reaches its apex with the realization of 50 kilims, in this case the work remains at the germinal state, that is, it is concluded in the linguistic medium. The consequence of such a status of absence of physical implementation is a specific pureness of *All numbers end in three* (hereinafter indicated by the acronym TINFIT, after the Italian title). Before investigating deeper the content of the statement that coincides with the work, a formal description of the process is appropriate.

Let's consider a function $y = f_{lettercounter}(x)$. The function is made up of (so to say) two submodules *cnv* and *cnt*: *cnv* converts a number into its linguistic representation, the *cnt* counts the characters composing the string resulting from *cnv*. The structure of $f_{lettercounter}$ is represented in Fig. 2, where number *m* is provided as input, converted into a string by *cnv*, its characters being counted by *cnt* and finally returned as *n*. The domain and the codomain of the function coincide with natural numbers ("the ones used for counting"). While the domain may eventually include zero, the image of the function on *y* cannot include zero: in fact, a number will be described by a string of at least 1 character. Thus, we will assume a stricter perspective, defining the domain and codomain as the positive integers.

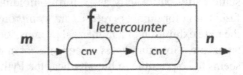

Fig. 2 $f_{lettercounter}$ with modules *cnv* (converter) and *cnt* (counter)

The values returned by $f_{lettercounter}$ for the first ten numbers are shown in Fig. 3: the x-axis represents the numbers, while the y-axis the number of letters that compose the relative Italian string[4]. The gray line indicates the values for which the number *n* is linguistically represented by a string containing *n* characters ($y = x$).

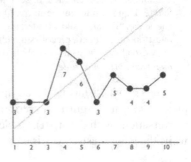

Fig. 3 $f_{lettercounter}$ for $n_{1...,10}$. The gray line indicates the values for which the number *n* is linguistically represented by a string containing *n* characters ($y = x$)

---

[4] As $f_{lettercounter}$ is defined only for natural numbers, the continuous curve has not a formal meaning (there is properly nothing between two adjacent numbers), but it is nonetheless useful for sake of visual clarity.

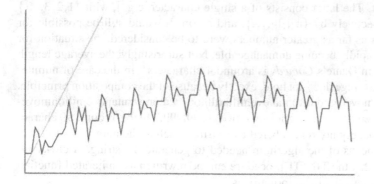

Fig. 4 $f_{lettercounter}$ for $n_{1...,100}$

6

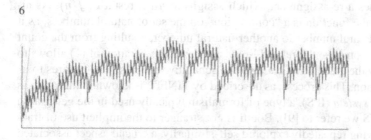

Fig. 5 $f_{lettercounter}$ for $n_{1...,1000}$

Fig. 4 and 5 represent the values of $f_{lettercounter}$ respectively for the first 100 and 1000 numbers, with the gray line increasing its slope only for graphical reasons.

It is easy to observe that the function shows a very irregular pattern for the first ten values, while a clear and consistent pattern emerges when greater numbers are taken into account. This nonlinearity can be explained by an empirical linguistic fact. The first ten numbers are encoded with entirely arbitrary strings (e.g. 1, 7, 9, "uno", "sette", "nove"). While for successive values, and in a systematic way above 20 ("venti"), a composition principle holds (22, 23, 24, ["ventidue", "ventitre", "ventiquattro"] ..., 123 ["centoventitre"], ..., 923 ["novencentoventitre"] ...). This systematic feature shapes the trend of the curve. The negative peaks occur when 0 is used in some positions, e.g. in the case of 100 ["cento"] vs. 103 ["centotre"] vs. 123 ["centoventitre"]). The only case where the line $y = x$ crosses the curve is for $x = 3$, while only in the range $[1...5]$ we have $y \geq x$, i.e. the linguistic representation has a number of characters equal or greater than the number it represents. Above 5, the ratio $\frac{y}{x}$ tends to zero. That is, y grows much less than x. Thus, while 3 is composed of 3 characters ("tre", $\frac{y}{x} = 1$), 99 ("novantanove") consists of only 11 characters ($\frac{y}{x} = 0.111...$). This makes indeed sense in terms of mnemonic economy: if the language were to follow the straight line $y = x$ then it would take 100 characters to encode the number 100, 200 characters to encode the number 200, and so on. The language would encode numbers as in the case of a unary alphabet. The latter consists of a single character, e.g. |, with $\{1,2,3,\dots\}$ represented respectively by $\{|,||,|||,\dots\}$, and so on. It would still be possible for small values, but as far as greater numbers were to be considered, the situation for memory would rapidly become unmanageable. Not surprisingly, the average length of Italian words in Dante's *Comedy* is around 4 characters[5]. In the case of number encoding, these average does not hold, exactly because of the composition principle. As an example, "novecentonovantanovemiliardinovecentonovantanovemilioninove-centonovantanovemilanovecentonovantanove" (999.999.999.999) counts 99 characters, and it is probably not remembered as a string labeling the number, but procedurally, i.e. by means of the algorithm needed to generate the string. A closer inspection reveals that the TINFIT procedure can be rewritten as an iterated function starting from $f_{lettercounter}$, i.e. informally as

$$n \mapsto f_{lettercounter}(n) \qquad (1)$$

where $\mapsto$ indicates a re-assignment, which assigns to n the result of $f(n)$. As already discussed, the function is a "contraction" on the set of natural numbers, as it associates each natural number to another natural number, resulting from the count of the letters that compose the related linguistic item. The notation of (1) allows to describe not only the association but also the generative process, that requires to iterate the association. This process, as described by TINFIT, is known in literature as *iterated function system* (IFS), a type of formalism typically used in the generation of fractals (on IFS we refer to [9]). Boetti is not stranger to the implicit use of fractals, the artist having repeatedly explored self-similarity, a crucial aspect associated with fractals. As an example, in *Afghanistan* (1980, see [1], 168) the shape of the map of Afghanistan is constructed with smaller shapes of the same map. Indeed, in the *Alternando* series, the larger square is made up of 100 squares, in turn made up of 100 squares. In relation to Fig. 2, the IFS can be represented as in Fig. 6. In the Boetti-Salerno formulation there is an exit condition for iteration (the "ending in three" clause), when $m = n$.

Fig. 6 clearly shows the feedback mechanism, whereby the output value is reinjected in the function input. In the IFS parlance the sequence of values thus obtained is called the "orbit" of the function. An orbit has a "fixed point" for $n = f(n)$: once n is reached, the system keeps on outputting n. The fixed point is also called the "attractor" of the system.

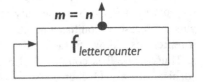

Fig. 6 $f_{lettercounter}$ as an IFS, with n feeding back the function

---

[5] To be precise, 4.045, calculated from the electronic version provided by Project Gutenberg (http://www.gutenberg.org/cache/epub/1012/pg1012.txt).

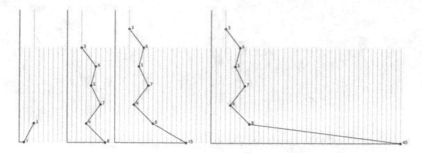

Fig. 7 Iterative $f_{lettercounter}$. Orbits for $n = 1, 8, 15, 40$

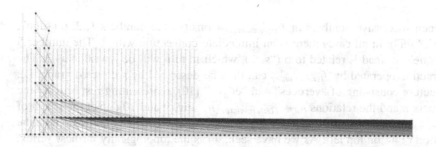

Fig. 8 Iterative $f_{lettercounter}$. Orbits for $n = 1, \ldots, 1000$

In the case of TINFIT, the system built on $f_{lettercounter}$ has clearly a fixed point, namely, 3, and the title of the work is exactly the description of the presence of the attractor. Fig. 7 consider the orbits of the system in relation to the number of iterations, that is, in terms of steps that lead to the attractor 3. The starting number is represented on the x-axis and the process develops along the y-axis. The iterations are represented by segments joining points: thus, a single iteration for $n = 1$ ("uno" counts 3 letters), five for $n = 8$, six for $n = 15$ and $n = 40$. Fig. 8 depicts the overlapping orbits for $n = 1, \ldots, 1000$ and shows how the convergence towards the attractor requires few iterations also for higher values of n: with $n = 1, \ldots, 1000$, up to a maximum of 8 in the case of $n = 44$ ("quarantaquattro"), whose orbit is $[44, 15, 8, 4, 7, 5, 6, 3]$.

3 Not really all the numbers end in three

However, some structural aspects of the system do not clearly emerge from the graphic representations used so far. Indeed, it is possible to observe that the orbits rapidly converge to a fixed set of values. This behavior is more evident if a different, more compact representation is considered, that takes into account the relationship

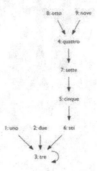

Fig. 9 Graph of $f_{lettercounter}$
for $n = 1, \ldots 9$

between successive numbers in $f_{lettercounter}$. Consider the numbers 1, 2, 6 ("uno", "due", "sei"): in all cases there is an immediate connection with 3. The number 5 ("cinque") instead is related to 6 ("sei"), which in turn will be rewritten as 3. The contraction operated by $f_{lettercounter}$ can then be described by a directed graph, i.e. a structure consisting of "vertices" and "edges" [10], representing respectively the numbers n and the relations $n \mapsto f_{lettercounter}(n)$. Fig. 9 and 10 depict the graph of $f_{lettercounter}$ for $n = 1, \ldots 9$ and $n = 1, \ldots 50$ respectively. The latter allows to note the location of 44 on top left. As we have seen, 44 is the topologically farthest vertex, as it takes 8 steps to get to the attractor 3 following the directed edges (the arrows). The graph provides a compact representation of the system of relations among the interested numbers, and can be thought as a generative grammar for sequences of numbers. The generation process can take any vertex as a starting point and then follows the path on the graph until a looping vertex is reached, that is, the attractor 3 that is connected only to itself. A path on the graph is another way of describing an orbit of the function. However, graph modeling allows to study some aspects of TINFIT that would be substantially hidden in the previous representations.

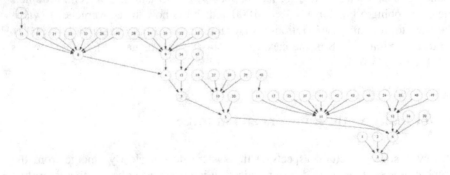

Fig. 10 Graph of $f_{lettercounter}$ for $n = 1, \ldots 50$, only numbers are represented for sake of clarity

In particular, is it really true that all the numbers end in three? Under which conditions? As Salerno said, it is only a matter of alternating numbers and letters, to read and to count". Now, this figure of alternation, that articulates the notion of double as a process, is at the core of the dual nature of TINFIT, a playful and at the same time icy work, and a real conceptual prototype of Boetti's *modus operandi*, holding together structure and history by alternating them at each step. Indeed, the number represents a cognitive primitive related to the idea of quantity. Counting is a cognitive performance in some way universal, that has no trade with cultural history. On the contrary, the letters that compose a number are the result of a triple cultural mediation: first of all, numbering systems encode quantity in different way in different cultures; then, different languages can encode even the same numbering system in different ways; finally, different writing systems and practices (including orthography) can encode even the same language in different ways. From this perspective, TINFIT takes as its basis –on the side of the "letters"– respectively the decimal positional system, its expression in the Italian language, and the Italian standard orthography. Thus, the previous formalization is related to a totally empirical and historical set of facts. Indeed, there is no formal necessity in the construction of strings that represent numbers in different languages, and the same composition principle that we have discussed before is just an empirical regularity historically attested in some specific languages. Hidden in the original formulation by Boetti and Salerno, this culturally related issue has immediately prompted in the discussion above, as the mixed Italian/English context probably forced the reader to note that in English all the numbers end in four (Fig. 11), and, just to take into account the narrow field of the Indo-European written languages, the same applies to German (Fig. 12). However, a lower style variant of *Rechtschreibung* (the German standard orthography) makes use of digrams instead of umlauts e.g. 'ö", "ä" are written as "oe", "ae". The resulting graph in Fig. 13 shows a remarkable new phenomenon, a topological variant: the system features not one but two attractors, 4 and 5, splitting the graph into into two subgraphs.

And yet, other topologies are possible. Fig. 14 represents the graph of $f_{lettecounter}$ for French, with $n = 1, \ldots, 20$. Not only the numbers do not end in 3, but they do not end at all, as the graph does not converge on a looping vertex, but on a circuit connecting $5 \mapsto 4 \mapsto 6 \mapsto 3 \mapsto 5$. As a consequence, there is not an attractor, that

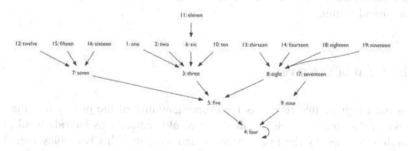

Fig. 11 Graph of $f_{lettercounter}$ for $n = 1, \ldots 20$, English

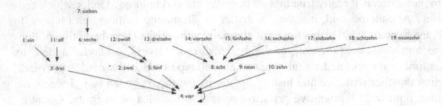

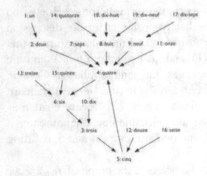

Fig. 12 Graph of $f_{lettercounter}$ for $n = 1, \ldots 20$, German

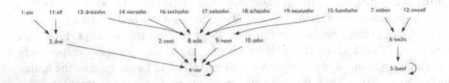

Fig. 13 Graph of $f_{lettercounter}$ for $n = 1, \ldots 20$, German with digrams

Fig. 14 Graph of $f_{lettercounter}$ for $n = 1, \ldots 20$, French

is, a vertex satisfying $n = f(n)$. The presence of an attractor, too, is therefore an entirely empirical matter.

4 On the form of the rivers

Thus, it is the language that re-injects the evenementiality of the history into the abstractness of the structure, which otherwise would remain - as Derrida would say - a totality deserted by the forces. History and structure: this two poles alternate and intermingle endlessly while reading the numbers and counting the letters.

Hence the purity of TINFIT as a Boettian prototype: a work without support that still links the formality of the function to the materiality of language, or to say it with a famous Boettian couple, the order of the number to the disorder of history. The most general feature, apparently independent from cultural parameterizations, is probably the limit to zero of the ratio $\frac{y}{x}$, precisely for reasons related to memory performance. It is for anthropological and phenomenological reasons that typically (though not always) the first ten numbers receive a name in the languages, as the result of the somatic constitution of the hand[6]. The infinite series of numbers and the infinity discreteness of language originate and converge at the same time in the "small quantity" of the fingers. Boetti has provided an admirable application of this duality between the number and the world, between order and disorder, in one of his most famous works, created in collaboration with Anne-Marie Sauzeau, *Classifying the Thousand Longest Rivers in the World*, a book resulting from an extensive research that progressively lists the World's 1000 longest rivers, each one with their basic geographical data. Here "the crazy bet" consists in "applying the classification to water" [11]. With an inversion that, however, retains the same terms, in *Classifying* the number represents a cultural order imposed on the complexity of nature. Not by chance, Boetti noted:

rivers are liquid, there is water, the female element, they move, there is the time.

But in the Platonic list of *Classifying* the number seems to lose its Pythagorean connotation, as a form of the becoming, so evident in *Alternando* and in TINFIT. Yet, is there a "form of the river"? While the empirical shape of the river depends on many factors, the (usually) convergent branching structure can be modeled through a fractal methodology [12]. The series of TINFIT, no matter how far it departs, gradually converges, step by step, to a single element. Put it in another way, each number is the source of a stream that gradually connects to a main course. In Italian, each number is therefore a tributary of 3, of a different order: a tributary of the tributary, a tributary of the tributary of the tributary, and so on. A main course, his few major tributaries, a series of tributaries that grows. Remarkably, the form that Boetti seals in the progression of TINFIT is thus the form of the river. The infinite numbers are the infinite streams that converge on the lower-order tributaries. In hydrology, the value representing the branching order for a tree structure is known as the Strahlen Stream Number. Fig. 15 represent the basin of the Amazon river (the longest reported in Boetti's book) as a graph connecting each river with its tributaries. While most British rivers have a Strahlen number between 3 or 5, the Amazon is the only river reaching 12[7].

In conclusion, even in its minimalism, the little conceptual work of TINFIT shows different aspects of Boetti's aesthetics linking together works that are apparently very different, above all through a common attitude to procedurality.

---

[6] Boetti explicits the relationship between the hand and the decimal system in *Postali 80*, [1], 51, where the sequence 1-10 is achieved through a system of contours of the fingers, as usually drawn on paper with pencil.

[7] http://en.wikipedia.org/wiki/Strahler_Stream_Order

138   A. Valle

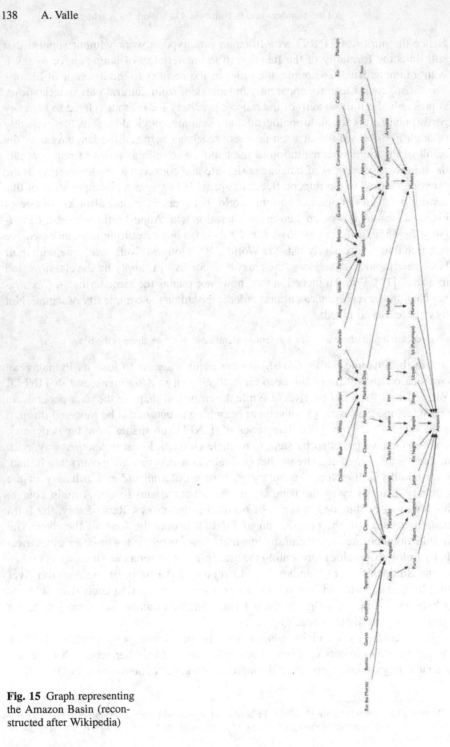

Fig. 15 Graph representing
the Amazon Basin (recon-
structed after Wikipedia)

Not by chance, rivers prompt Boetti to formulate a general reflection on ethics, precisely by taking into consideration their typical topology:

> The river has two senses. If you go upstream, at a certain moment you will have to choose, when you find a river crossing [...] you will face a crossroads, and after you will have to choose again. Whereas if you go along with the current towards the sea, you do not ever choose. Strangely enough, you will have two different attitudes using the same path[8].

References

1. J.C. Ammam, M.T. Roberto, A.M. Sauzeau (eds.), Alighiero Boetti 1965-1994. Mazzotta, Milano, 1996.
2. G. Celant, Arte povera. Galleria de' Foscherari, Bologna, 1968.
3. M. Margozzi, Alternando da 1 a zero e viceversa, 1993. In: S. Pinto (ed.), Alighiero e Boetti – l'opera ultima. SACS, Torino New York London Venezia. 1996, chap. 1, pp 21-32.
4. A. Valle, V. Lombardo, H. Vogel, Alternating from 1 to X and vice versa. In: ACM Multimedia (2007), pp. 922-931.
5. A. Valle, Scrivere con la sinistra è disegnare. Su grafie e notazioni. Il verri 38, 83, 2008.
6. G. Di Pietrantonio, C. Levi (eds.), Alighiero Boetti – Quasi tutto. Silvana, Milano, 2004.
7. P. Zellini, Gnomon. Adelphi, Milano, 1999.
8. G.B. Salerno, Arte della copia e misteri della riproduzione. In: J.C. Amman, M.T. Roberto, A.M. Sauzeau (ed.), Alighiero Boetti 1965-1994. Mazzotta, Milano, 1996, pp. 47-53.
9. M.F. Barnsley, Fractals everywhere, 2nd edn. Morgan Kaufmann, San Francisco, 1993.
10. R. Diestel, Graph Theory. Springer, New York, 2000.
11. A. Sauzeau, E. Greco, Shaman showman Alighiero e Boetti – Niente da vedere niente da nascondere. Luca Sossella, Roma, 2006.
12. I. Rodrìguez Iturbe, A. Rinaldo, Fractal River Basins. Chance and Self-Organization. Cambridge University Press, Cambridge, 1997.

---

[8] The two excerpts are from a video interview by Antonia Mulas, from the series *In prima persona. Pittori e scultori italiani* (1984).

Mathematics and Literature

The Unreasonable Effectiveness of Mathematics in Human Sciences: the Attribution of Texts to Antonio Gramsci

Dario Benedetto, Emanuele Caglioti and Mirko Degli Esposti

1 *Style*, where are you?

From the online Oxford vocabulary we can read the following definition:
[style] *noun*

- a particular procedure by which something is done; a manner or way;
- a way of painting, writing, composing, building, etc., characteristic of a particular period, place, person, or movement;
- a way of using language;
- a distinctive appearance, typically determined by the principles according to which something is designed: a particular design of clothing.

Does it sound strange that mathematicians are interested in this? As we try to argue, it is probably not completely silly that certain topics are faced with pure or almost pure mathematical eyes. *Creativity* means *generation* of novel, original and, hopefully, coherent structures from the use of elementary elements and with the use of old or newly created rules. This is, of course, a very general and debatable definition, but one that works not only for literature, but also for music, painting and any other form of artistic creativity.

But *generation of structures* is exactly what mathematicians usually do and study (well, roughly speaking) and we think that some of the aspects related to creativity, style (whatever it means) and the like can be modeled, quantified and sometime measured and simulated. Of course, we are researchers and we must be very careful to not abandon too quickly the safe shore of a rigorous scientific method in trying to catch the *phlogiston* of style and creativity. Asserting that a given author has a unique and detectable style that can be measured in any of his creations is of course not only a poor scientific statement, but also completely wrong in its generality.

Dario Benedetto, Emanuele Caglioti
Department of Mathematics, Sapienza University of Rome (Italy).
Mirko Degli Esposti
Department of Mathematics, University of Bologna (Italy).

Emmer M. (Ed.): Imagine Math. Between Culture and Mathematics
DOI 10.1007/978-88-470-2427-4_14, © Springer-Verlag Italia 2012

This is why we say that, as for the concept of race, *style does not exist*. But still in any creative process, nothing is really generated from scratch and any process of content creation (a piece of a text, for example) is always the result of a complex interaction between the author's experience and skills from the past, what the author has created up to now, the topic of the content, the author's desire for originality, and much more of course. Because of this, it is not completely foolish to imagine that some patterns characterizing the author's style might be hidden in the created work. These features are probably not always sufficient to identify and discriminate the author with respect the rest of the universe, but are probably sufficient to identify and distinguish the author within a coherent and limited framework made of a few properly selected authors and topics.

More precisely, in case of written texts, even if a general definition of style does not means anything and probably it does not even exists, we believe that certain abstract quantities, a little bit more general than the *usual semantic or syntactic* (e.g. words) structures, sometimes contains useful information that allows use to *discriminate* the real author of the text among a finite set of possible authors. This is basically the aim of *Authorship Attribution* (often denoted by A.A), a quite old area, a field where philology, computer science and (we believe) pure mathematics and physics come together.

Again with the aim of being general and, even more, generic, let us start with a very simple example in *visual attribution*. Look, for example, at the fragments of picture in Fig. 1: two of them are from Henry de Toulouse-Lautrec and the other two from Pierre-Auguste Renoir. Our guess is that it will take you just a small fraction of a second to *clusterize* them with respect the author.

Can we do it with an algorithm? Well, in this case it should not be difficult to recognize some very natural *features*, that can be extracted from each image and makes it possible to discriminate between the artists. Here it would be enough to extract very general and simple information about the statistical distribution of small geometrical features, such as segments or circles and arcs. But things might be a little bit more complicated: think about being able to distinguish a set of drawings securely attributed to the great Flemish artist Pieter Bruegel the Elder (1525-1569) from a set of imitation Bruegels, whose attribution is generally accepted among art historians. Here the task is much more difficult, even for humans, and was attacked

Fig. 1 A simple problem in *visual attribution* [16]

only very recently with quite sophisticated mathematical tools, such as wavelets and *sparse coding* [5].

Computer science has in fact been grappling with the very general problem of clustering and discriminating objects of different nature for very long time now and a very fundamental type of approach has been developed in the last decades: first of all, define and extract suitable features and then, in the spirit of machine learning, train and use a suitable algorithm/machine (neural nets, supported vector machine, Bayesian tools, etc...) to discriminate, classify and clusterize objects into several known or unknown classes.

Here, as already stated, we would like to concentrate on literary texts and the the task of using quantitative tools, either statistical or more purely mathematical, to attribute a given anonymous or apocryphal text to a specific author. This area of research has a quite long history but we can trace one of its crucial and fundamental steps back to the work of the physicist Thomas Corwin Mendenhall (1841-1924). As described and discussed in [6], in the article "The characteristic curves of composition" [8] T. C. Mendenhall was attracted by the similarity between the statistical distribution of words of various lengths (how many words of length 1, 2, 3, and so on) and the spectrum generated by the spectroscopic analysis, a very innovative and much discussed technique in the last decades of the nineteenth century. In fact, as he wrote in the 1887 issue of *Science* (see [6] and references therein):

> It is proposed to analyse a composition by forming what may be called a "word spectrum" or "characteristic curve" which shall be a graphic representation of the arrangement of words according to their length and the relative frequency of their occurrence.

In 1901 T.C. Mendenhall published in *The Popular Science Monthly* an article with the title *A Mechanical Solution of a Literary Problem* [7] where he studied and compared body of works by Shakespeare, Marlowe and Bacon facing the already classical open question regarding the real identity of the author of the literary works traditionally transmitted under the name of Shakespeare. The aim of Mendenhall was to verify whether the style of Shakespeare's works was either unique and detectable or if it was similar (if not identical) to the style expressed in the work of Marlowe or Bacon, under the (wrong) assumption that the frequency distribution of word lengths was a unique *finger print* of the author's style (we now know that this simple assumption is unfortunately very far from true).

In this contribution, following [1] and in particular [2], we want to describe the method used to attribute articles whose author was unknown to Antonio Gramsci (a famous politician, philosopher and journalist, who was one of the founders of

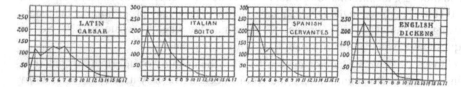

Fig. 2 Word spectra traced by Thomas Corwin Mendenhall in [8]

the Italian Communist Party in the 1921), which we have developed together with M. Lana [6]. The techniques we used are not the mere result of experiments; on the contrary they are based on some important ideas of modern mathematics that we believe are useful for distinguishing this kind of research from the substantially empirical approach generally used in this field, as we show in the following section.

2 The measure of information content

Information theory was born in 1948 with the article [10] by Claude E. Shannon "A Mathematical Theory of Communication", which poses and solves the problem of defining the amount of information contained in a "message", for example a text or more generally any sequence of symbols (for a more extended account see the book of Pierce [9]).

The unit of measurement of information is the *bit* ("binary unit"), it is the measure of the information which chooses one of the two elements of an alternative: on/off, open/closed, right/wrong, true/false, 0/1 (which are the two symbols used in the binary numeration system). With one bit available you can make just two distinct assertions; with two bits, four "words" can be made (in binary digits the four words are: 00, 01, 10, 11); with three bits, eight words can be made, and so on. The amount of information corresponding to eight bits is called *byte*; with one byte you can generate 256 different words. With 256 possibilities, an entire alphabet of a western language can be codified; indeed the letters (including capitals, accented letters, punctuation marks, special symbols) are never more than 256. Each letter is therefore represented by a sequence of eight bits, through universally accepted "codings" (like ASCII and iso_8859-15).

An example: the DNA sequence AGCTTTTCATTCTGACTGCA is composed of 20 characters and a text file containing it is 20 byte large. One could therefore think that this sequence contains $20 \times 8 = 160$ bits of information. Actually to write a DNA sequence an alphabet consisting of the 4 letters A, C, G, T is sufficient, and since you can codify 4 letters with just 2 bits, the given sequence contains (at most) $20 \times 2 = 40$ bits of information.

So the coding affects the quantity of information used to write a message: while 8 bits per character are used for an English text, for a "genetic text" 2 bits per character are enough. One can imagine the strangest codings: for the sequence

```
TTTTTTTTTTTTTTTTTTTTTTTTTTTTTTTTTTTTTTTTTTTTTTTTTT
```

(50 times T), the information content is 400 bits if it is codified in the Latin alphabet, 100 bits if codified in the DNA alphabet, a few bytes in any programming language, using a command which corresponds in human language to "write 50 T's". So the question is: what is the information size of the sequence?

In his 1948 work Shannon determined that the quantity of information contained in a message is the minimum number of bits needed to codify it, and defined *entropy* as the minimum number of bits per character. There are programs which attempt to codify a message using the least possible number of bits: they are the data compression programs (for example WinZip on Windows OS, gzip and bzip2 on Unix OS; for a general description of compressors see for example [12]). The compression rate (obtained by comparing the dimension of the compressed text with that of the original text) allows us estimate the entropy of a text (see also [13]). Shannon's theory has a rigorous and consistent formulation only for well-defined mathematical objects, but mathematicians find it natural to use his ideas in the field of text analysis as well: one can indeed make the hypothesis that, by measuring the compression rate of an author's texts, an intrinsic quantity of it is measured. Shannon himself, through an experiment, estimated that the the average quantity of information of the "English language" is between 0.6 and 1.3 bits per character. Though the entropic characteristics of an author's writing are certainly interesting, they are not very useful for attribution problems.

By developing Shannon's ideas, one can obtain an effective instrument for the attribution problem: the concept of *relative entropy*. In order to illustrate this concept it is useful to analyse in detail how some methods (algorithms) for data compression work. Those which were discovered first (Shannon-Fano, Huffman) use a priori knowledge of the character statistics of the text and codify just one character (or a few characters) at a time. The more frequent the character is used, the shorter the code assigned to it. As an example, consider Morse code, which, even though it is not a compression code, has been conceived to meet a similar requirement: speeding up the transmission of messages in the English language. Morse code uses 5 characters: line, dot and short break to codify letters, medium break to separate words, long break to separate sentences. The more frequently used letters in English are codified with a shorter sequence, and so transmitted faster: "e" is codified with "." while "z" is codified with "–..". The letter frequency distribution was studied a priori: Morse visited a print shop to obtain it. The entropy is the minimal number of bits per character needed to codify a sequence, so if the sequence is codified in a non-optimal way more bits than necessary are used; given two sequences, their relative entropy is precisely the number of bits per character which are added when one sequence is codified with the code which is optimal for the other sequence. The example of Morse code is useful to understand this concept. Let us suppose that Morse code is optimal for the English language: if it is used to codify an Italian message it gives a text longer than the one which would have been obtained by using a Morse code optimized for the Italian language. The difference of length (per character) is a measure of the relative entropy between English and Italian.

Relative entropy is a very powerful tool to quantify the difference among sequences, and therefore among authors. As early as 1993 Ziv and Merhav in [15] had proposed the use of relative entropy to deal with problems of categorization and suggested definite methods to measure it. These methods have been proposed and used on more specific problems in the fields of biological sequence analysis and of authorship attribution (see the survey of Stamatatos [11]).

In particular, we have used a method based on the ideas of the compression algorithm LZ77 of Ziv and Lempel [14], which is the origin of the compression softwares zip/WinZip/gzip (see [3] for an explanation of the method you can easily implement!). For example, as an estimate of the relative entropies of the tragedy *Oedipus at Colonus* by Sophocles and the tragedy *Alcestis* by Euripides with respect to *Antigone* by Sophocles, we obtain the following values:

$$D(\text{Oedipus at Colonus}||\text{Antigone}) = 0.130, \quad D(\text{Alcestis}||\text{Antigone}) = 0.244$$

In this example the relative entropy of two texts by Sophocles is lower than that between a text of Euripides and a text of Sophocles.

3 The statistics in the texts

Assuming that the text is "just" a symbol sequence means not taking into consideration either the content of the text or its grammatical aspects: letters of the alphabet, punctuation marks, blank spaces between words are just abstract symbols, without a hierarchy. Moreover, as a basic constituent of the text, the word has no more meaning than other aggregates of symbols, and its role as a unit of higher level than the single character is described by the n-*gram*. Here are some useful examples:

- by *monogram* (1-gram) we mean one single symbol of the alphabet;
- by *bigram* (2-gram) we mean a sequence of two symbols, for example "me" but also also "e " (i.e. "e" followed by a blank space);
- by *trigram* (3-gram) we mean a sequence of three symbols, for example "the", but also "e.L";
- by n-*gram* we mean any sequence of *n* symbols; for example "the entr" is an 8-gram.

From a simplifying mathematical point of view, besides considering the text as an abstract sequence of symbols, we also assume that it has been *generated by a source*, symbol after symbol. The nature of the source is not the object of analysis, but is only an abstract model of all the entities which can generate texts. The source emits its messages (texts) choosing the symbol to be emitted each time according to probabilistic rules. The difference between sources is due to the different probabilistic rules used to generate the messages.

Obviously, this source/message scheme is too rigid and abstract to be a reasonable interpretation of the author/text relationship. In particular, in the mathematical models for symbol sources it is possible to make the rules for symbol generation explicit, while it is at least doubtful that such rules exist for a real author writing a text. On the other hand this approach gives some useful indications, as Shannon showed in constructing the *approximations* of the texts [10].

He defined the approximation of "order 0" simply by extracting symbols randomly, all with the same probability. Obviously the texts obtained in this way are far from resembling an English text, as can be seen from the following example:

```
pmR!.ALvPRW;sVfjyaicGlWsN;lDADdHWiCAWEF.cbLG;UgdPYCFbUGmH:eMiVtK
```

The "first order" approximation is obtained by extracting symbols with probabilities equal to the relative frequencies with which they are found in the reference corpus. An example of text obtained like this is:

```
orklpa tea yohhranKgoc suhoruhytenffari, ed e aelutnGb u.ifaaepn
```

With the "second order" approximation a significant difference is introduced: the new character is chosen in relation to its antecedent. For example, to choose the character following a c, we have to compute the frequencies of the bigrams beginning with c in the corpus and divide them by the overall frequency of c; the values obtained in this way are the *conditioned frequencies*. A text generated with such rules is for example:

```
he cerye Huro ut thowaverowolesthirliror me g imen andy lind f g
```

Correspondingly, a model of third order is obtained by measuring the frequencies of a character with respect to the two previous ones. An example is

```
at st the waid heithand by hinglittlyints napt th hothed that han
```

and an example of approximation of the 9th order is

```
and their guns across their wandering which would no more left
```

The number of the characteristics of the original texts which are preserved in the models grows with the order of approximation: in the first order approximation the separation into words is similar to the one of English language; in the second order the syllables are substantially correct, and the beginning and end of words are plausible; the ninth order approximation roughly respects grammatical rules.

With this idea in mind, we can suppose that the "stylistic" differences among authors *must* result in numerical differences between n-gram frequencies. In this way, once the n-gram frequencies of an unknown text are measured and compared with the "typical" n-gram frequencies of an author, it is possible to perform the attribution by choosing the author for which the difference between the known and calculated frequencies is minimal.

4 Mathematical methods for attribution

Using the description of texts as n-gram sequences and the entropy as a measure of information content, one can develop procedures for the attribution of a text whose author is unknown using two basical tools:

- for n-grams: their frequencies in the text are measured and compared with the ones of the available authors;
- for entropy: the relative entropy of the text in respect to the available authors is calculated.

Any mathematical method for attribution will be characterized, in brief, by the following two aspects:

1. choice of the "objects" for which the counting is significant, that is, of the objects which are supposed to be used with frequencies of occurrence significantly different from that of the other authors;
2. choice of the way of translating the measures of the quantities described in 1. into attributions.

The n-gram frequencies of occurrence and the relative entropy are two possible choices for the counted objects in 1. The most used methods for the choices in 2 are those involving probabilistic/statistical techniques and those of metrics or similarity ones. Probabilistic and statistical methods start from the basic assumption that the characteristics of a text (the ones chosen in point 1. are not univocally linked to an author, but occur with different frequencies of occurrence for the different authors). There are very well-established mathematical techniques (for example, Bayes' formula and more generally statistical tests) for the study of the inverse problem, that is to calculate the probability that a text with certain observed characteristics has been written by a given author.

A different approach consists in synthesizing as a single quantity the difference/dissimilarity observed measuring the quantities chosen in point 1). This value will be a measure of the proximity of two texts or of a text and an author; in general, the number is lower when the measured difference is smaller, i.e., when the texts are closer to one another. At that point a mathematician will prefer to define this proximity as a "distance" (or "metric"), which is a definite mathematical concept, obtained by abstracting the characteristics of the usual distance between points in space [1].

There are two advantages with respect to a generic measure of "proximity" are two: "distance" is a mathematically solid, unambiguous concept, and it allows the use of other mathematical tools that have been developed using the notion of distance. It is important to remark that the metric description allows among other things the construction of "phylogenetic trees", where attribution corresponds to the inclusion in different branches of the tree (see e.g. [4]).

5 Gramsci or not Gramsci?

The basic ideas and considerations described up to now were further developed and implemented for the attribution of journal articles whose author was unknown but which were probably written by Antonio Gramsci (we did this work for the new "Edizione nazionale degli scritti di Antonio Gramsci", in collaboration with "Fondazione Istituto Gramsci Onlus"[2]) While we do not go here into the details of our attribution procedure (we refer the interested reader to [1] and [2] for further

---

[1] Specifically the distance satisfies the "triangular inequality", which in essence states that if going from A to B one deviates passing through C, the path becomes longer.

[2] http://www.fondazionegramsci.org/ag_edizione_nazionale.htm

mathematical and methodological details), we would like to give here just a brief overview of the fundamental steps, together with a sketchy graphical visualization of some achievements.

We started with a tuning stage, in which we selected two methods, one based on *n*-gram (with *n*=8) and the other based on entropic techniques, and we tested it measuring the distance between any given text X out of a group of 50 *Gramscian* and 50 known *non-Gramscian* texts and the other 99.

Having fixed a distance method (say the one based on *n*-grams) and having chosen text *X*, we now have 99 numbers representing the distances of *X* from the individual elements of the the corpus. This suggests two questions:

- What can we expect about the distance of *X* from the 49 texts of the class to which it belongs?
- Is it possible to consider conveniently all the information contained in these numbers, hence going beyond the simple (and inefficient) attribution given by the author of the *nearest text*?

With these questions in mind one can in fact define, for any given text *X* and for each distance method, a *Gramscianity index* $-1 \leq v(X) \leq 1$ calculated from the above 99 numbers;the index value will be greater the closer the unknown text is to the group of Gramscian texts. A value near to 1 (-1) provides a strong indication of a correct attribution to Gramsci (to non-Gramsci), while values near to 0 are a mark of great undecidability. The index allows a quite direct and useful graphical visualization of the attribution, as shown for the training set in Fig. 3 (see [1, 2] for more details).

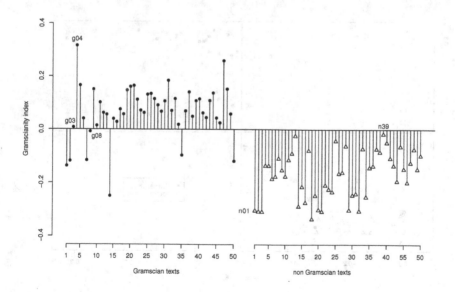

Fig. 3 Attribution of the 100 texts, using the Gramscianty index given by the n-grams method

We can now repeat the same procedure with the entropic distance, which is based upon completely different principles. On the other hand, one might fear that they in fact will give the same information, adding nothing to the accuracy of the global method. We have therefore made sure, with suitable methods, of the statistical independence of the rankings of the texts ordered following the two distances.

In the end, we have attributed to Gramsci only the texts that both methods assign to him. Moreover, both methods give a numerical value for the attribution, so that it is possible (and very useful) to give a two-dimensional graphical representation of the overall results: the Gramscianity index obtained with the n-gram method is plotted on the horizontal axis where positive values correspond to the attribution to Gramsci, negative values to "non-Gramsci". The rightmost points are the texts for which the attribution to Gramsci is more certain, the leftmost are those for which the method suggests with greatest certainty an attribution to authors other than Gramsci. On the vertical axis the value of the analogous index given by the relative entropy method is shown; in this case advancing from bottom up means moving from suggested non-Gramscian texts to suggested Gramscian ones.

The results of the tuning stage is shown in Fig. 4.

The first quadrant, therefore, contains the texts that both method attribute to Gramsci. In this case, for example, there is no triangular point among them, meaning that there are no false positives (no wrong attributions to Gramsci). The number of texts correctly attributed to Gramsci is 43, the 86% of the total. In the second quadrant lie the texts attributed to Gramsci by the relative entropy method but not by the n-grams method. There are no texts in the fourth quadrant: these would be those

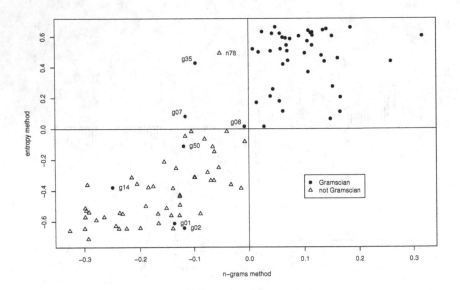

Fig. 4 Attributions for the 100 texts of the tuning stange

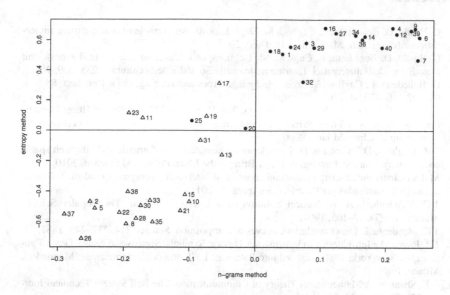

Fig. 5 Attributions for the 40 texts of the blind test

attributed to Gramsci by the *n*-gram method but not by the entropic method. Finally, the third quadrant contains the texts not attributed to Gramsci by either methods.

The second stage was a *blind test*, performed on 40 texts (attributions were known but communicated to us only *a posteriori*). The results we obtained is shown in Fig. 5. One can see that 18 Gramscian texts out of 20 are correctly attributed to Gramsci, which is the 90%, with no false positives.

This was just a very brief description of the beginning of the analysis, which was followed by a systematic attribution of thousands of actual unattributed articles. The attributions obtained with these mathematical methods are at present in the hands of experts of Gramsci, who are re-elaborating and re-discussing the results in light of a more traditional philological approach. We do not intend to go here into the details of this interesting (and regarding some aspects, original) interaction between mathematical based quantitative methods and philological methods, besides confirming a fruitful coherence between the approaches, which will be discussed elsewhere.

Of course, *it's a long way to the top*[3] and a lot of interesting questions are still not completely answered: for example, Why do these abstract methods work? Which *n*-grams really contain the *signature* of Gramsci's style?

To put it another way: *style, style, where are you?*

---

[3] ... *if You Wanna Rock 'n' Roll*: AC/DC, from the album *T.N.T.*

References

1. C. Basile, D. Benedetto, E. Caglioti, M. Degli Esposti, An example of mathematical authorship attribution. Jour. Math. Phys. **49**, 1–20, 2008.
2. C. Basile, D. Benedetto, E. Caglioti, M. Degli Esposti, L'attribuzione dei testi gramsciani: metodi e modelli matematici. La matematica nella Società e nella cultura **3**, 235–269, 2010.
3. D. Benedetto, E. Caglioti, V. Loreto, Language Trees and Zipping. Phys. Rev. Lett. **88** n. 4, 048702-1–048702-4, 2002.
4. L.L. Cavalli-Sforza, P. Menozzi, A. Piazza, The History and Geography of Human Genes. Princeton University Press, Princeton, 1994; translated in Italian as: Storia e geografia dei geni umani. Adelphi. Milano, 2000.
5. J.M. Hughesa, D.J. Graham, D.N. Rockmore, Quantification of artistic style through sparse coding analysis in the drawings of Pieter Bruegel the Elder. PNAS and Science, 2010.
6. M. Lana, Individuare scritti gramsciani anonimi all'interno di un corpus giornalistico. Il ruolo dei metodi quantitativi per l'attribuzione (preprint 2011).
7. T.C. Mendenhall, A Mechanical Solution of a Literary Problem. The Popular Science Monthly **LX**(7), 97–105, 1901.
8. T.C. Mendenhall, The characteristic curves of composition. Science **214**, 237–249, 1887.
9. J.R. Pierce, An Introduction to Information Theory: Symbols, Signals and Noise. Dover Publications, New York, 1980; traslated into Italian as: La Teoria dell'Informazione. Mondadori, Milano, 1983.
10. C.E. Shannon, A Mathematical Theory of Communication. The Bell System Technical Journal **27**, 379–423, 623–656, 1948.
11. E. Stamatatos, A Survey of Modern Authorship Attribution Methods. Jour. Am. Soc. Infor. Sci. Tech. **60**(3), 538–556, 2009.
12. I. H. Witten, A. Moffat, T.C. Bell, Managing Gigabytes, 2nd edition. Morgan Kaufmann Publishers, 1999.
13. A.D. Wyner, Typical sequences and all that: Entropy, Pattern Matching and Data Compression. Shannon Lecture, IEEE Information Theory Society Newsletter, 1995.
14. J. Ziv, A. Lempel, A universal algorithm for sequential data compression. IEEE Transactions on Information Theory **IT-23**(3), 337–343, 1977.
15. J. Ziv, N. Merhav, A measure of relative entropy between individual sequences with application to universal classification. IEEE Transactions of Information Theory **39**(4), 1270–1279, 1993.
16. Wikimedia Commons, su http://commons.wikimedia.org/wiki/:
File:Pierre-Auguste_Renoir_-_Study_for_'Dance_in_the_Country',
_pencil,_1883.jpg Honolulu Academy of Arts;
File:Portrait_de_Suzanne_Valadon_par_Henri_de_Toulouse-Lautrec.jpg;
File:Henri_de_Toulouse-Lautrec_053.jpg The Yorck Project: 10.000 Meisterwerke der Malerei. DVD-ROM, 2002. ISBN 3936122202. Distributed by DIRECTMEDIA Publishing GmbH.;
File_Renoir_-_Sitzendes_M

Lost in a Good Book:
Jorge Borges' Inescapable Labyrinth

William Goldbloom Bloch

1 Entrance

"Entrance" is a marvelous example of a heteronym, a word endowed with (at least) two separate pronunciations, each of which has a meaning distinct from the others. In this case, in the context of what I am writing about, the two pronunciations and meanings complement each other very well: Entrance, a place to enter, and: Entrance, to delight and fill with wonder.

Many might say that Fiction is the art of putting together words to evoke images that, in turn, provoke feelings and touch beliefs in the readers. Many might also say that Mathematics is the art of assembling axioms and instantiations of Platonic Forms to express that which Must Necessarily Be So. These descriptions make the prospect of combining Mathematics and Fiction akin to the workings of a cinematic mad scientist, wildly pouring chemicals together, boiling and distilling them, and hoping that an improbable outcome will occur (some critics might allege that all of Mathematics is Dreary and Boring Fiction, but I do not imagine I am able to convince them of the error of their ways).

It is difficult to envision how the two fields may be brought together in a way that does justice to the artistry, the precision, and the profundity of these human projects. The Argentine writer, Jorge Luis Borges, is one of the earliest, and one of the finest writers to do so. Borges was born in Buenos Aires in 1899, spent time in Switzerland and Spain during and after World War I, then moved back to his beloved Buenos Aires in 1921. During the 1920s and the early 1930s, he gained notice, in Argentina at least, for his manifestos on poetry, his poetry, his minor fiction, and for his essays that appeared in unusual venues, such as the Argentine version of the Ladies' Home Journal. When Borges was approaching his 40th birthday, it would have been easy to dismiss him as a failure: He was unmarried, living with his mother, and was working in a menial capacity at a small local library in Buenos Aires, dreaming of the days

William Goldbloom Bloch
Wheaton College, Norton MA (USA).

Emmer M. (Ed.): Imagine Math. Between Culture and Mathematics
DOI 10.1007/978-88-470-2427-4_15, © Springer-Verlag Italia 2012

when gauchos and vaqueros strode the grassy plains of Argentina and were as minor deities.

During this period of his life, as he was slowly going blind, along with many eclectic and varied offerings at the National Library of Argentina, Borges was voraciously reading and rereading books by Bertrand Russell on the philosophy and practice of mathematics. His searches and researches were actuated, as were mine initially, by the desire to understand infinity in both the large and in the small; in the large in the form of the vastness of the space time continuum, particularly as seen in Nietzsche's meditation on eternal recurrence,[1] and in the small in the form of the increasingly tiny steps taken by both Achilles and the Tortoise in Zeno's famous paradox.[2]

In no way should we consider Borges a mathematician, nor would he have called himself one, either, but by virtue of his immersion in the ideas of the combinations of atoms, and the manifestations of infinity in the large and small, and in Russell's resolutions of Zeno's paradoxes and musings on transfinite numbers, Borges transformed himself into a sort of receptacle of the ideas of the mathematics of permutations and of the infinite. His ruminations about these ideas and how they played out in the world around him, finally found expression in *Ficciones*, a slim collection of remarkable short stories that account for the beginning of his literary fame outside of Argentina.

It is possible, and very tempting, to look to the vast, increasing, swampy river-delta of literary works and figures, and pigeonhole Borges as a successor to the gothic inventions of Edgar Allan Poe, as the inventor of magic realism, and as a precursor to many of the writers of science fiction, metafiction, and hyperfiction. It is equally possible, and maybe even more tempting, to reference his unique and memorable prose style; Nobelists Gabriel García Márquez and Maria Vargas Llosa, and other eminent writers in Spanish such as Carlos Fuentes and Julio Cortázar, all cite Borges as a profound influence.

But I come not to bury Borges with praise, nor as an exegete, nor as a literary analyst or theorist. For that matter, I am not even going to address the most obvious incursions of mathematics in his short stories.[3] Rather, the rest of this paper focuses on the largest conceivable labyrinth in one of his Ficciones, the sublime story "La Biblioteca de Babel." The idea of the Library of Babel is based on simple combinatorial mathematics, and Borges admits to expanding it from a perfunctory science fiction story, Kurd Lasswitz's "Die Universalbibliothek" (1901).

---

[1] If time is unlimited, and the Universe is bounded, Nietzsche argued that given the existence of only a finite number of atoms, all events must eventually recur infinitely often. A similar argument has been made by those who postulate that if the Universe is unlimited spatially, then infinitely many exact replicas of this moment of you reading this footnote are simultaneously distributed throughout the Universe.

[2] Zeno claims that the fastest, Achilles, may never catch the slowest, the Tortoise, for by the time Achilles has caught up to where the Tortoise was, to Achilles' dismay, the Tortoise has moved on, and Achilles must chase him further yet.

[3] There is a whole Universe of approaches I am not taking with this article, and while it may be an interesting exercise to catalogue more of them, including the one in which I misunderstand him to be a South African writer who mysteriously writes in Spanish and avoids discussion of serious racial and ethnic identity issues, there are only so many footnotes available to me.

2 The Unlimited Labyrinth

In a nutshell: Imagine an alphabet consisting of 25 symbols, including the period, the comma, and the blank space. The Library consists of all books containing every single possible sequence of those symbols contained in 410 pages, where each page has 40 lines, and each line has 80 symbols. As Borges writes:

> The Universe (which others call the Library) is composed of an indefinite, perhaps infinite number of hexagonal galleries. In the center of each gallery is a ventilation shaft, bounded by a low railing. From any hexagon one can see the floors above and below - one after another, endlessly. The arrangement of the galleries is always the same: Twenty bookshelves, five to each side, line four of the hexagon's six sides; the height of the bookshelves, floor to ceiling, is hardly greater than the height of a normal librarian. One of the hexagon's free sides opens onto a narrow sort of vestibule, which in turn opens onto another gallery, identical to the first - identical in fact to all. [...] Each wall of each hexagon is furnished with five bookshelves; each bookshelf holds thirty-two books identical in format; each book contains four hundred ten pages; each page, forty lines; each line, eighty black letters.[4]

The joy of Borges' story is that in a few paragraphs, he creates an entire universe dedicated to holding the many, many books of the Library,[5] to evoke the miasmic atmosphere of an all-encompassing Library, and to educe metaphysical, moral, and psychological consequences of residence in such an uncompromising milieu. The narrator of the story is an aged Librarian, who has spent a lifetime speculating on the meanings and ways of the unfathomable Library, and a question that puzzles him is similar to one that puzzles us: What is the shape, the configuration, of the vast edifice? Early in the story, narrating as the Librarian, Borges writes:

> Let it suffice for the moment that I repeat the classic dictum: The Library is a sphere whose exact center is any hexagon and whose circumference is unattainable.[6]

Then, after many meditations, the final sentences of the story invite us to reopen the question of the topology of the Library:

> I am perhaps misled by old age and fear, but I suspect that the human species - the only human species - teeters at the verge of extinction, yet that the Library - enlightened, solitary, infinite, perfectly unmoving, armed with precious volumes, pointless, incorruptible, and secret - will endure.

> I have just written the word "infinite." I have not included that adjective of out of mere rhetorical habit; I hereby state that it is not illogical to think that the world is infinite. Those who believe it to have limits hypothesize that in some remote place or places the corridors and stairs and hexagons may, inconceivably, end - which is absurd. And yet those who picture the world as unlimited forget that the number of possible books is not. I will be bold enough to suggest this solution to the ancient problem: The Library is unlimited but

---

[4] From "The Library of Babel," pages 112-113, as translated by Andrew Hurley in *Collected Fictions*.

[5] $25^{1,312,000}$ books, enough to completely fill $10^{1,834,013}$ universes the size of our own (using the current best estimate). To gain intuition about these numbers, and for many other insights and perspectives, see: *The Unimaginable Mathematics of Borges' Library of Babel* [1].

[6] Ibid, p. 113.

periodic. If an eternal voyager should journey it in any direction, he would find after untold centuries that the same volumes are repeated in the same disorder - which, repeated, becomes order: The Order. My solitude is cheered by that elegant hope.[7]

As satisfying as it may be to the Librarian to invoke infinity in the large, and have a Universe unfettered by walls or boundaries, most of our minds balk at that sort of forever. (To say nothing of the construction costs or upkeep!) Furthermore, one must also ask, "What does a hexagon rest on?" The only possible answer is the hexagon below it, which leads to the next question, "And what does that hexagon rest on?" This evokes the old joke that the earth rests on a elephant, and the elephant on a tortoise, and then when curious about what the tortoise rests on, the punch line emerges: It's all tortoises from there on down. If the construction truly is "all hexagons from there on down," the Library is certainly worse than a castle built on sand. Rather remarkably, the architectural model of the Library proposed below provides a satisfying answer to this question.

A note regarding the gravity of the situation. If the Universe and the Library are synonymous, and if we make the reasonable assumption that the Universe is neither expanding nor contracting, it follows that the natural gravitational field would be identically zero everywhere. Even though there are vast amounts of matter in the Universe/Library, its homogeneous distribution entails that the gravitational effect from any one direction would be canceled out by precisely the same effect from the opposite direction. Since the builders of a Library must be, at least from our perspective, omnipotent, their talents surely must include the ability of imposing a useful constant gravitational field on the Library.

Even with these constraints (or allowances), is it possible, then, to find an elegant path through the maze that is the Universe that is the Library? In other words, is it possible to find an appropriate description for the Universe that is the Library that best satisfies the Librarian's perspectives while not violating the intuition that infinite is too big?

Topology is a branch of mathematics that explores properties and invariants of spaces, and for the purposes of this paper a space will be considered as a set of points unified by a description. It is not unreasonable to speculate as to a topology of the Library that best reflects the anonymous Librarian's received wisdom and secret hopes.

Collecting from above the properties of the classic dictum (CD) and the Librarian's solution (LS):

1. The Library is spherical. (CD)
2. The center can be anywhere - there is uniform symmetry. (CD)
3. The circumference is unattainable. (CD)
4. There are no boundaries. (LS)
5. The Library is limitless. (LS)
6. The Library is periodic. (LS)

---

[7] Ibid, p. 118.

Is there a space that embodies all six of these properties? If so, how can it be best envisioned and grasped by the intellect? As it turns out, there is an excellent candidate that almost perfectly addresses these properties.

To understand the candidate space, it is reasonable to begin with the space most familiar to the intuitive geometric sense: Euclidean three-dimensional space (henceforth, 3-space). It is a space that possesses volume, it has three axes of orientation with ourselves as the central point, and we may move forward or backwards, we may move left or right, and we may move up or down. And, of course, we may also move in combinations of these directions. Notice that from this description, there is no fixed preferred center point: We are our own central points.

Indeed, one of Descartes' deepest ideas was to specify a point - some point, any point - in 3-space and call it the origin. Three axes intersecting at the origin, typically called the x-, y-, and z-axes, are set with each axis positioned at right angles to the other two. They abstract our innate, intuitive orientation and, with the introduction of a unit length, which naturally induces a numbering of the axes, give rise to a coordinatization of space.

But there are no distinguished points of any kind in Euclidean 3-space; in fact, the view from any point is the same as from any other point. There are no walls, no boundaries, and no limits. At the end of the story the Librarian envisioned this kind of space, partitioned into hexagons, filled with books, extending infinitely throughout the totality of 3-space. The books' shelving pattern repeats endlessly along each of the three axes, much as a symmetric wallpaper pattern does in two dimensions. While this conception of the Library satisfies points (2), (4), (5), and (6) from the list above, it also induces a vertiginous disorientation described above.

So it is seen that Euclidean 3-space embodies some of the qualities of interest in the quest to understand the large-scale structure of the Library. It is necessary to sketch two more ideas, one mathematical, one mystical, before describing a substructure for the Library that better reconciles the characteristics of the Classic Dictum and the Librarian's Solution.

The mathematical idea is relatively recent - it comes from the early part of the twentieth century. For the purposes of this paper, it suffices to say that a manifold is a space that is locally Euclidean, but that on a global scale may be non-Euclidean. Perhaps the simplest possible example is that of a sphere, or the surface of the earth. Locally, assuming that we are so small we can't detect the curvature, each micropatch of a sphere is, in essence, a two-dimensional Euclidean plane (2-space). Think of the steppes of Central Asia, of the corn belt of the United States, of the Sahara desert, or of any large, calm body of water to engage vivid testimony on this point. Globally, despite the essential flatness of each little patch, there is non-Euclidean behavior: If we begin at a point, pick a direction, and continue moving in that direction, we circumscribe the sphere and return to our starting point. This can't occur in 2-space, where we travel forever in one direction and do not ever come close again to a previously visited point.

Again, a manifold is locally Euclidean. If we start at any point in space, look around and take a few steps in any direction, do we think we are in Euclidean space? If the answer is yes, then we are in a manifold. If we continue walking, and some unusual phenomenon occurs, such as returning to our starting point, then we realize we are in a nontrivial manifold; that is, one with global non-Euclidean properties. Our universe, for example, seems to be a manifold, although interesting questions arise at black holes. Certainly one cannot imagine standing at a black hole and taking a step in any direction!

The mystical idea is relatively ancient - I leave it to a Borgesian intellect to trace its roots and age-long echoes. Begin in a familiar place, our own universe. If we talk about an object in our universe - for example, a tomato, or a desk, or a chair - we view it as embedded in a larger space.

Consequently, we often use our relative coordinate system to refer to objects, as when we say "It's on my right," or "Over there! Directly behind you, to the left," or "Scratch my back . . . lower . . . lower . . . to the right . . . now up . . . that's it!" Over the millennia, primarily as navigation aids, we've settled upon somewhat less arbitrary reference points, such as the North Star, the magnetic North Pole, and the true North Pole. The point is, though, that these references, these origins, are all within our universe. "Outside the universe" is a phrase beyond normal comprehension. Some theories place our universe in a larger matrix, such as a superheated gas cloud containing an infinite number of inaccessible universes, or in a higher-dimensional space, or in a multiplicitous welter of bifurcating universes. However, these theories inevitably (should) raise the question,

What is outside of the larger universe?

The answer is no thing; nothing; non-space; indescribability; un-thing-ed-ness; Void beyond Vacuum: All these non-things are the "outside" of the universe. These two ideas, the mathematical and the mystical, are woven together in this question and its answer.

Where is the center of a sphere?

If the sphere is considered as an everyday object embedded in 3-space, the answer may take a form such as "at the intersection of two diameters," or, pointing at it dramatically, saying with particular emphasis, "There! In the middle, in the interior!"

If, though, we consider the surface of the sphere as a manifold, as a space in itself and of itself, then the question and answer are subtler. As in the case of our universe, as if we were points residing in the sphere itself, there is no legitimate referral to a point outside the universe of the surface of the sphere. There is only the sphere; everything else is no-thing. Where is the center of a sphere? Considered as a manifold the answer is

Everywhere and nowhere.

Every point has the property that locally, it looks like Euclidean space, and regardless of the direction taken, consistently moving in any chosen direction returns an intrepid traveler to the starting point. No point is distinguished in any way.

One more idea is necessary to provide a satisfying topology for the Library. The example of a manifold used above is a two-dimensional sphere (2-sphere). There are a number of ways to rigorously define a 2-sphere. Euclid might write something

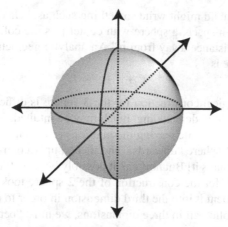

Fig. 1 In which the coordinate axes in Euclidean 3-space are seen to intersect at a point that is inside, but not part of, the 2-sphere

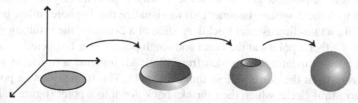

Fig. 2 A disk curls up out of Euclidean 2-space, and when the boundary contracts to a point, a 2-sphere is formed

like, "Given a point p in 3-space, a sphere with center p is the collection of all points a specified uniform distance away from p." An analytic geometric equation for the standard unit sphere is

$$x^2 + y^2 + z^2 = 1.$$

In Fig. 2, using words and pictures, is a topological construction of a 2-sphere.

Start with a disk in the Euclidean plane, and while preserving the interior of the disk except for bending and stretching, crimp the entire boundary circle up out of 2-space, and then contract the boundary to one point. This point, the contraction of the boundary, becomes the north pole and vanishes into the surface of the sphere created as the process is completed.

An interesting point: The way it's been described, and the way the picture shows this process, it may seem as though a disk is being modified over time. By contrast, though, one should simply say, "Identify the boundary of the disk to a point." Thus, in some sense, the creation of the sphere is a timeless step that happens "all at once."

The three-dimensional sphere (3-sphere) provoked many advances in topology over the past century, and due to the recently solved Poincaré conjecture, it remains a vibrant research topic. The 3-sphere is a generalization of the 1-sphere - a circle

- and the 2-sphere. Euclid might write something such as, "Given a point p in four-dimensional Euclidean space, a sphere with center p is the collection of all points a specified uniform distance away from p." An analytic geometric equation for the standard unit 3-sphere is

$$w^2+x^2+y^2+z^2= 1.$$

An analogous topological construction for the 3-sphere is difficult to envision, but by pushing the limits of understanding, much may be intuited.

Take a solid ball and, while leaving the interior of the ball uncompressed, crimp the entire boundary 2-sphere "upwards," and then simply contract the boundary 2-sphere to one point. That's it! But how can it possibly be done? At least the difficulty is easy to understand, for the construction of the 2-sphere took a two-dimensional object, the disk, and bent it into the third dimension in order to contract the boundary. Starting with a solid ball in three dimensions, we must "bend" the ball into the fourth dimension before we can contract the boundary.

At this juncture, the mathematics becomes unimaginable; the best to be hoped for is that by meditating on the lower-dimensional examples accessible to the imagination, it may be possible gain some sense of what is possible. By proceeding from analogies with the 2-sphere, two methods to visualize the 3-sphere follow below.

If we take a two-dimensional Euclidean slice of a 2-sphere, the resulting geometric object is either a point - at the north and south poles - or a 1-sphere.

Using a mild updating of an idea from Flatland, imagine a movie of a planar slice moving from the north pole to the south pole. The frames show a point that grows into a unit circle, which then shrinks back down to a point (figure 4, left). In a similar fashion, imagine taking a three-dimensional Euclidean slice of a 3-sphere. In that case, the resulting geometric object is either a point - at the "top" or "bottom" - or a 2-sphere. If we make a movie of the voluminous slice moving from the top to the bottom, a viewer would see a point that grows into a unit sphere, and then shrinks back down to a point (Fig. 4, right).

Expanding this idea, suppose we were forced to squish the 2-sphere, whose natural home is in 3-space, down into 2-space. Since the 2-sphere can be conceived as a collection of stacked circles combined with two poles, it may be envisioned as

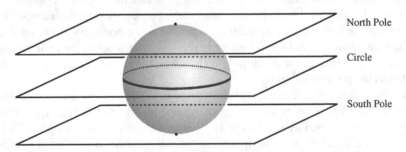

Fig. 3 Planar slices of a 2-sphere yield a single point at the north and south poles, and a circle everywhere else

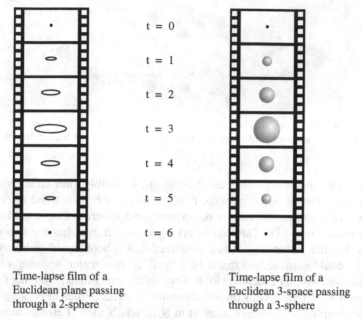

Time-lapse film of a
Euclidean plane passing
through a 2-sphere

Time-lapse film of a
Euclidean 3-space passing
through a 3-sphere

Fig. 4 On the left side, planar slices are taken from the north pole of a 2-sphere to the bottom, while on the right side, analogously, a volume slice is passed from the "north pole" of a 3-sphere to the "south pole"

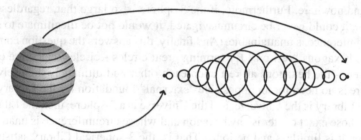

Fig. 5 A representation of a flattened 2-sphere in the Euclidean plane as an infinite collection of circles plus the two points of the poles

a flattened planar depiction of a collection of intersecting circles with two points signifying the north and south poles.

The related problem is how to represent the 3-sphere after it has been down-dimensioned into 3-space. By thinking of the 3-sphere as "stacked" 2-spheres - in the same sense that a 2-sphere is stacked 1-spheres - the analogous 3-space representation is a collection of intersecting 2-spheres.

All the girders and struts of the framework are now in place to finish assembling the topology and cosmology of the Universe that is the Library. The 3-sphere is a three-dimensional manifold; at every point of the 3-sphere, an inhabitant would say – locally - that space is Euclidean. If a group of travelers walked what was perceived

Fig. 6 An analogous operation is performed on a 3-sphere, in which it is "flattened" out into Euclidean 3-space to an infinite collection of spheres along with the two points of the poles

to be a straight line in any direction, they would - possibly after countless ages - return to their starting point; as such, the 3-sphere can be construed as periodic. There are no boundaries, no walls to bump into; the 3-sphere is limitless. Moreover, in his luminous story "The Garden of Forking Paths," Borges has the sympathetic sinologist Stephen Albert say, "I had wondered how a book could be infinite. The only way I could surmise was that it be a cyclical, or circular, volume, a volume whose last page would be identical to the first, so that one might go on indefinitely." Even though Albert rejects this line of reasoning for "The Garden of Forking Paths," this quote, coupled with Borges' interest in Nietzsche's idea of eternal recurrence, indicates that Borges was willing to consider cyclic or recurrent structures as tokens of, or synonymous with, infinity.

Considered as a three-dimensional manifold, the center of the 3-sphere is everywhere and nowhere. Furthermore, if the 3-sphere is so large that, regardless of our transport, it could never be circumnavigated, it would not be illegitimate to say that the circumference is unattainable. (And finally, this answers the question concerning what the hexagons "rest on." By forming great circles - circles which are equators of a sphere - the hexagons all rest upon each other and ultimately themselves, and thus there is no need for an impossible "external" foundation for the Library[8]).

If the Library is the Universe, and the Universe is a 3-sphere, then the Library is a sphere whose exact center is any hexagon and whose circumference is unattainable; moreover, it is limitless and periodic. That is, the 3-spherical Library satisfies both the classic dictum and the Librarian's cherished hope.

3 Exit

In a classical maze or labyrinth, there is one entrance and one exit. In the novice's conception of mathematics, a problem has exactly one entrance, which is the problem itself, and the problem has exactly one exit, precisely that of the correct so-

---

[8] The spherical "volume slices" from figure 6 encourage the viewpoint that a 3-sphere is a collection of two points and infinitely many 2-spheres, which is, perhaps, the easiest way to visualize it. However, in 1931 the great mathematical theorist Heinz Hopf developed a way to fiber the 3-sphere into a collection of infinitely many 1-spheres (circles); that is, every point of the 3-sphere lies on one of infinitely many unit circles. There are no isolated poles in this decomposition of 3-sphere.

lution. A more experienced practitioner of mathematics knows that good problems have many points of embarkation and many ways of solving the puzzle, successfully leaving it in the past. Indeed, the more entrances and the more resolutions, the richer the implications of the problem, and the more likely it will be to resonate in the Sphere of Mathematics beyond itself.

In Borges' Library, and in Borges' conception of our Universe, the construct is that of a maze of interconnected layered labyrinths, and that form suggests a single entrance paired with a single exit. But Borges helped lead us beyond classical conceptions while still respecting them. The Library of his story is a labyrinth synonymous with the Universe. There are simultaneously no entrances, yet many entrances, for any entity capable of spawning life provides entrance for another being into the Library.

Similarly, there are no exits from the Library that is the Universe, and simultaneously many exits, although each such one partakes of the same stark fact of death. It is possible that such an exit allows entrance into a world of clarity that leaves all labyrinths behind, but that is not given for us to know, nor for Borges' fictitious Librarian to know, either.

From the outside, looking in, Mathematics provides a way to finesse the idea of infinity in the large, and see the Library as an enormous 3-sphere, a manifold of size beyond comprehension, a way to organize a cosmological substructure for the many books. From inside the story, looking out, the fictional Librarian knows just enough mathematics to understand that a profound joke has been played on him, but not enough to laugh at it.

References

1. A.M. Barrenechea, Borges: the Labyrinth Maker. Translated by Robert Lima. New York University Press, 1965.
2. W.G. Bloch, The Unimaginable Mathematics of Borges' Library of Babel. Oxford University Press, New York, 2008.
3. J.L. Borges, Ficciones. Emecé Editores, Buenos Aires, 1944.
4. J.L. Borges, Collected Fictions. Translated by Andrew Hurley. Penguin Classics, New York, 1998.
5. K. Lasswitz, The Universal Library. Translated by W. Ley in: C. Fadiman (ed.), Fantasia Mathematica. Copernicus, New York, 1997.
6. J. Munkres, Topology. Prentice-Hall, New Jersey, 1975.
7. J. Weeks, The Shape of Space: How to Visualize Surfaces and Three-Dimensional, Manifolds. Dekker, New York, 1985.

Mathematics and Applications

The Many Faces of Lorenz Knots

Marco Abate

One of the greatest pleasures in doing mathematics (and one of the surest signs of being onto something really relevant) is discovering that two apparently completely unrelated objects actually are one and the same thing. This is what Étienne Ghys, of the École Normale Superieure de Lyon, did a few years ago (see [1] for the technical details), showing that the class of Lorenz knots, pertaining to the theory of chaotic dynamical systems and ordinary differential equations, and the class of modular knots, pertaining to the theory of 2-dimensional lattices and to number theory, coincide. In this short note we shall try to explain what Lorenz and modular knots are, and to give a hint of why they are the same. See also [2] for a more detailed but still accessible presentation, containing the beautiful pictures and animations prepared by Jos Leys [3], a digital artist, to illustrate Ghys' results.

1 What is a knot?

Informally speaking, a knot is a closed piece of string in space. More formally, a *knot* is a (globally injective) embedding of the circumference S^1 in the Euclidean 3-space $\mathbf{R}^3$. Two knots are considered the same if there is a way of continuously deform the space $\mathbf{R}^3$ so to bring the first knot exactly onto the second knot (or its mirror image). In more technical terms, two knots are *equivalent* if there is a homeomorphism of $\mathbf{R}^3$ (a bijective continuous transformation of the space onto itself with a continuous inverse) transforming the first knot in the second. In particular, a knot equivalent to the standard unit circumference in the plane is actually unknotted, and thus considered a trivial knot. See, e.g., [4] for a not exceedingly technical introduction to the mathematical theory of knots.

Mathematicians are entomologists at heart; they are prone to uncontrollable classification urges. For instance, one would like to have a list of all possible knots (up to equivalence, of course). The usual way for representing a knot consists in projecting it onto a plane so that the projection crosses itself in a finite number of points, and only two strands of the knot pass through any crossing point. So one may look for the projection with the least number of self-crossings of a given knot (or, more precisely, of all equivalent forms of a given knot), and try to organize the knots

Marco Abate
Department of Mathemathics, University of Pisa (Italy).

Emmer M. (Ed.): Imagine Math. Between Culture and Mathematics
DOI 10.1007/978-88-470-2427-4_16, © Springer-Verlag Italia 2012

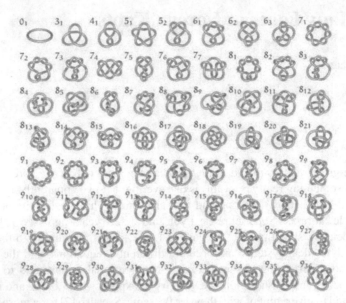

Fig. 1 Knots (modified from www.knotplot.com)

according to this least number of self-intersections. For instance, the trivial knot clearly admits a representation with no self-crossings: the standard circumference. It is not difficult to see that knots admitting representations with only one or two self-crossings are actually unknots; so the first non-trivial knot is the *trefoil knot,* whose representation (see knot 3_1 in Fig. 1) has exactly three self-crossings. Fig. 1 contains representations of all distinct knots with at most 9 self-crossings.

A particular subclass of knots will be useful later on. A *torus* is a doughnut-shaped surface, that is the Cartesian product of two circumferences; a *torus knot* is a knot on a torus. In other words, in a torus knot the string winds on the surface of a torus. Fig. 2 contains the representations of the simplest torus knots; see [5] for more pictures of knots.

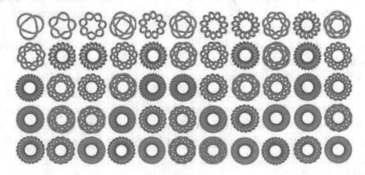

Fig. 2 Torus knots (modified from www.knotplot.com)

2 What is a Lorenz knot?

Lorenz knots appear in the first, and still most famous, example of chaotic dynamical system, introduced by Edward N. Lorenz in 1963 [6] as a simplified model for convection in the atmosphere. This model consists of three (mildly non linear) ordinary differential equations:

$$x' = 10(y - x), \quad y' = x(28 - z) - y, \quad z' = xy - (8/3)z.$$

How Lorenz noticed the presence of chaos in this system is by now almost legendary. He was solving numerically this system on a (large, for the time) computer, but he had to interrupt the computations for the night. The next day he gave as input to the computer the results of the computations of the previous afternoon, and soon noticed that the results he was obtaining were sensibly different from the ones he got the day before, even though the initial conditions were the same. Or were they? After several weeks of careful checking of the programs and computers involved to rule out any possible mistake, Lorenz realized that the data he entered the second day were only *approximations* of the data stored into the computer; and even though they were very good approximations (to the sixth decimal digit or so), this apparently negligible difference at the beginning provoked hugely different outcomes at the end.

Lorenz had discovered one of the most distinctive characteristics of chaotic dynamical systems: *sensitive dependence on initial conditions*. The slightest change in the initial state can cause a completely different result, the so-called (and by now exceedingly famous) butterfly effect. But in his model Lorenz also discovered another butterfly, which is more relevant to the present discussion.

The Lorenz model, as any system of ordinary differential equations in three variables, prescribes at each point in space a velocity vector; we can then start from any point in space, and move according to the speed and direction given by these velocity vectors. The itinerary we follow is an *orbit* of the model. Lorenz noticed that almost all orbits tended to accumulate onto a peculiar and approximately butterfly-shaped set, having a very intricate geometric structure (later on it was proved that it is a fractal set of dimension slightly larger than two). This set, the *Lorenz attractor*, was the first example of strange attractor for a chaotic dynamical system; check [3] for beautiful pictures of the Lorenz attractor, and [7] to play with different orbits and see in real time how they accumulate onto the Lorenz attractor (and how they depend on the choice of initial conditions).

Most orbits go around wildly getting closer and closer to the Lorenz attractor; but a few special ones actually lives in the Lorenz attractor itself. These are the *periodic* orbits: orbits that after a finite amount of time come back to their starting point. Periodic orbits are thus (never self-intersecting) closed curves in Euclidean space, that is, they are knots. And yes, the *Lorenz knots* are exactly the periodic orbits of the Lorenz model.

It turns out that Lorenz knots fill out (they are dense, another typical feature of chaotic dynamical systems is the coexistence of periodic behavior with very wild behavior) the Lorenz attractor, and so understanding them might give important information on the structure of the Lorenz attractor. In the Eighties Joan Birman and Bob Williams [8] started studying Lorenz knots, trying to understand and classify them. They showed that all torus knots are Lorenz knots; and very recently Birman and Ilya Kofman have proved that every Lorenz knot is a *twisted* torus knot, a knot that can be obtained from a torus knot by a simple procedure (amounting to cutting the knot in several carefully chosen places, twisting the strands according to specific rules, and then gluing the strands back together; see [9] for details).

3 What is a modular knot?

To explain what is a modular knot we must first explain what a lattice is.

Roughly speaking, a *lattice* is a discrete family of points (in a line, a plane, a space) uniformly distributed. The easiest example of lattice is the set of integer numbers in the real line; and, in a sense, this is the only example of lattice in the line. Indeed, if we take any lattice in the line, up to a translation we can assume that it contains the origin; and up to a rescaling we can assume that it is *normalized*, that is that the distance between two consecutive points in the lattice is exactly one - and thus we have recovered the integers. From a geometrical point of view, then, a lattice in the line is obtained by covering the line with infinitely many copies of the same basic block, an interval (of length one if the lattice is normalized).

In the plane, the situation is considerably more complex. As building block for a lattice we can use a parallelogram; but even assuming (as we may up to a translation) that one of the vertices of the parallelogram is the origin, we still have infinitely many distinct cases to consider. If one vertex is the origin, to describe the parallelogram (and hence the lattice obtained by covering the plane with copies of the basic parallelogram) it suffices to give the coordinates of two other vertices, $v_1 = (a_1, b_1)$ and $v_2 = (a_2, b_2)$. Furthermore, we can also assume that (up to a rescaling) the lattice is *normalized*, that is that the basic parallelogram has area one (conditions amounting to requiring that $a_1 b_2 - a_2 b_1$ is equal to one).

So to describe a normalized lattice we need four real numbers (the coordinates of two vertices of the basic parallelogram) satisfying one condition (area equal to 1); this means that we can identify the space of all normalized lattices with a suitable subset of the Euclidean 3-space (actually one needs to add a point at infinity, getting a subset of the 3-dimensional sphere, but this is a detail). It turns out that this subset is exactly the complement of a trefoil knot – the first but not last appearance of knots in this setting.

There is another way of describing the space of normalized lattices. Instead of considering the two vertices separately, we can put their coordinates in a 2x2 matrix; the normalization condition then amounts to saying that the determinant of this matrix is 1. If we multiply a matrix with determinant 1 by another matrix with determinant 1 we still get a matrix with determinant 1, that is another normalized lattice. In particular, this holds if we multiply by the diagonal matrix having e^t and e^{-t} as diagonal elements, where t is any real number. Letting t vary in the real numbers, we then get a whole family of normalized lattices, that can be thought of as a curve in the complement of the trefoil knot, an orbit of the *modular flow*. See [10] for (a lot) more details.

The modular flow appears and is very important in several areas of number theory and one-dimensional complex analysis; but the aspect that is interesting for us now is that the modular flow has periodic orbits, forming knots contained in the complement of the trefoil knot; these periodic orbits are (of course) called *modular knots*. It turns out that they are in one-to-one correspondence with (similarity classes of) 2x2 matrices with integer coefficients, determinant one and absolute value of the trace (the sum of the diagonal elements) greater than 2; these matrices are the *hyperbolic* elements of the *modular group* (the group of 2x2 matrices with integer coefficients and determinant 1). Notice that to give a modular knots it then suffices to give four integer numbers (satisfying a bunch of conditions); so it is not surprising that topological properties of modular knots have something to do with number theoretical properties of integer numbers.

Modular knots have been studied for a long time; however, Ghys found a new way of looking at them, giving unexpected results.

4 What do they have to do with each other?

Ghys' surprising discovery is that a knot can be realized as a Lorenz knot if and only if it can be realized as a modular knot. In other words, the class of Lorenz knots coincide with the class of modular knots.

To prove this, Ghys gave a way to pass from a Lorenz knot to a modular knot and conversely, based on the idea of Lorenz template previously introduced by Birman and Williams. The Lorenz template (see again [2] and [3] for beautiful pictures) is a figure-eight-shaped surface, similar to - and thus still sort of butterfly-like - but much simpler than the Lorenz attractor, with the very useful property that every Lorenz knot can be continuously pushed onto the Lorenz template (remaining equivalent to the original knot). Furthermore, the left wing and the right wing of the butterfly in the Lorenz template are joined by a central one-dimensional stick; and every Lorenz knot must cross this central stick. More precisely, Birman and Williams showed that a Lorenz knot is completely determined by the way it crosses the central stick, going into the left wing or the right wing after each crossing; the sequence of left/right choices is enough to completely reconstruct the given Lorenz knot.

What Ghys did was to find a (topologically equivalent) copy of the Lorenz template inside the space of normalized lattices (the complement of the trefoil knot), and to show how modular knots can be (following a natural geometric procedure) pushed down on this Lorenz template so to become Lorenz knots. Conversely, he also showed that every sequence of left/right choices at the central stick can be realized by a modular knot, and so all Lorenz knots are modular knots too.

This discovery has already had profound consequences in the theory of the modular flow (and thus in number theory and related areas). All properties of Lorenz knots must be enjoyed by modular knots, and conversely. For instance, modular knots must be fibered (that is, it should be possible to fill the complement of the knot by surfaces all having the boundary lying on the given knot, quite an unusual property for a knot to have) because (as Birman and Williams showed) all Lorenz knots are; at present a direct proof (that is a proof not using Lorenz knots) of this fact is not known.

Another unexpected consequence consists in a new way to compute the Rademacher function, a very useful number-theoretic object whose classical definition is very cumbersome, involving taking the complex logarithm of the 24th root of something known as the Weierstrass discriminant, and then following the complex logarithm along a closed curve associated to a (hyperbolic) element of the modular group. Going along a closed curve the complex logarithm changes by an integer multiple of $2\pi i$; this integer is the value of the Rademacher function computed in the given element of the modular group. Well, Ghys has shown that the Rademacher function is simply given by the number of the left choices minus the number of right choices made by the corresponding modular knot pressed onto the Lorenz template!

Ghys' discovery prompted new advances in the study of the Lorenz model too; for instance, the characterization of Lorenz (and hence modular) knots as twisted torus knots given by Birman and Kofman was inspired by Ghys' results. Furthermore, modular knots are much easier to generate than Lorenz knots, and since they still preserve all the topological features of Lorenz knots, in principle they might be used to explore the intricacies of the Lorenz attractor. In general, the appearance of important features of the Lorenz model in a completely different context seems to indicate that it was not a complete accident that the first chaotic system to be discovered was Lorenz'; possibly the Lorenz model is more basic, more intrinsic than we actually imagine. This is probably just the beginning of a long and exciting story: new discoveries, new results and new unexpected connections might be waiting just around the corner. But even if this will not be the case, Ghys' work remains a beautiful piece of contemporary mathematics that will be studied and admired for a long time.

References

1. É. Ghys, Knots and dynamics. In: International Congress of Mathematicians. Vol. I: European Mathematical Society, Zürich, pp. 247–277, 2007.
2. http://www.josleys.com/articles/ams_article/Lorenz3.htm
3. http://www.josleys.com/
4. C. Adams, The knot book. American Mathematical Society, Providence RI, 2004.
5. http://www.knotplot.com
6. E. Lorenz, Deterministic nonperiodic flow. J. Atmos. Sci. 20, 130-141, 1963.
7. http://www.cmp.caltech.edu/~mcc/Chaos_Course/Lesson1/Demo8.html
8. J. Birman, R. Williams, Knotted periodic orbits in dynamical systems. I. Lorenz's equations. Topology 22, 47–82, 1983.
9. J. Birman, I. Kofman, A new twist on Lorenz links. J. Topology 2, 227–248, 2009.
10. J-P. Serre, A course in arithmetic. Springer-Verlag, Heidelberg, 1973.

Waiting for ABRACADABRA.
Occurrences of Words and Leading Numbers

Emilio De Santis and Fabio Spizzichino

In this paper we introduce the readers to the concept of "leading number", as proposed by J. H. Conway in the seventies of the last century. The leading number, associated to a word **w**, is a binary vector that describes some special aspects of the structure of **w**. We shall see that it conveys the essential information that is needed in the analysis of the time of occurrence of **w** in a random sequence of letters.

The theme of time of occurrence of words, a sort of classical topic of applied probability, presents several aspects of interest. In particular, it gives rise to some apparently paradoxical conclusions. Furthermore it is related with the notion of fair games and leads to interesting mathematical problems.

1 Words and Leading Numbers

Here we fix a (finite) set containing N elements $a_1, ..., a_N$:

$$\mathscr{A} \equiv \{a_1, ..., a_N\}.$$

For our purposes, $\mathscr{A}$ will be called an *Alphabet* and its elements will also be termed as *letters*.

An ordered sequence $\mathbf{w} \equiv w_1 w_2 ... w_k$, where any of the elements w_j is one of the letters taken from $\mathscr{A}$, will be called a *word of length k on the alphabet $\mathscr{A}$*.

Here are at once some simple examples:

a) if $N = 26$ and $a_1 = A, a_2 = B, ..., a_{26} = Z$ then $\mathscr{A}$ is the alphabet of the English language and $\mathbf{w}_1 \equiv BLUE$, $\mathbf{w}_2 \equiv PINK$ are two different words of length 4 on $\mathscr{A}$;

b) if $N = 2$ and $a_1 = 0, a_2 = 1$, a word of length k on $\mathscr{A} \equiv \{0, 1\}$ can be seen as an integer number M written in the binary system ($0 \leq M \leq 2^{k+1} - 1$, with $M = 0$ if $w_1 = w_2 = ... = w_k = 0$ and $M = 2^{k+1} - 1$ if $w_1 = w_2 = ... = w_k = 1$);

c) whenever $N = 10$ and $\mathscr{A} \equiv \{0, 1, ..., 9\}$, a word on $\mathscr{A}$ can be seen as an integer number written in the ordinary decimal system;

Emilio De Santis and Fabio Spizzichino
Department of Mathematics, Sapienza University of Rome (Italy).

Emmer M. (Ed.): Imagine Math. Between Culture and Mathematics
DOI 10.1007/978-88-470-2427-4_17, © Springer-Verlag Italia 2012

d) when R stands for *Red* and B stands for *Black,* and 0 for *zero*, a sequence of outcomes, such as *BRRR0BRRRB0BR*, at a Casino's roulette, can be seen as a word of length 13 on the alphabet $\mathscr{A} \equiv \{R,B,0\}$ with $N = 3$.

We stop here, but of course a lot of further pertinent examples could be produced. Before continuing, some almost obvious, but useful, remarks are also in order:

α) for a fixed alphabet $\mathscr{A} \equiv \{a_1,...,a_N\}$, and for a length k, any choice $w_1 w_2...w_k$ is admissible. No other structure, possibly giving rise to special constrains, is considered; in other terms, there is no *dictionary* to be respected. Notice for instance that, like *BLUE* and *PINK*, also *CIAO* or *TTTT* are considered as words of four letters on the English alphabet, even if they do not belong to the dictionary of English language;

β) it can happen that a same symbol can represent *letters* from two, or more, different *alphabets*. For instance 0 can be seen as the ordinary integer number zero or the special outcome of a roulette;

γ) a same real object might be represented by means of two (or more) different words on two different alphabets;

δ) repetitions of letters are allowed in a word, so that we can admit the case $k > N$.

We can say more about item δ): repetitions of pattern of letters in a word are important in our context, as it will be explained in the next section. Our interest there will just be concentrated on some special effects of such repetitions in random sequences of letters from an alphabet.

For our purposes, and in order to describe particular aspects of the pattern of repetitions in a word **w**, it is interesting to recall the concept of leading number, as introduced by Conway (see in particular the citation in [3]).

Let $\mathbf{w} \equiv w_1 w_2...w_k$ be a word of length k on the alphabet $\mathscr{A}$. Then we define as *leading number* associated to **w**, the k-coordinates *binary* vector

$$\varepsilon_{\mathbf{w}} \equiv (\varepsilon_{\mathbf{w}}(1), \varepsilon_{\mathbf{w}}(2),\ldots,\varepsilon_{\mathbf{w}}(k)),$$

where each $\varepsilon_{\mathbf{w}}(u)$ is equal to 0 or to 1, according to the following position: for $u = 1,2,\ldots,k$

$$\varepsilon_{\mathbf{w}}(u) = \mathbf{1}\{w_{k-u+1}\ldots w_k = w_1 \ldots w_u\},$$

that is

1. $\varepsilon_{\mathbf{w}}(u) = 1$ if the *sub-word* made with the *first u* letters of **w** coincides with the *sub-word* made with the *last u* letters of **w**.
2. $\varepsilon_{\mathbf{w}}(u) = 0$ otherwise. Notice that two words are different if there is at least one *position* at which they present two different letters.

Example 1. Let $k = 11$ and let **w** be the word *ABRACADABRA*, a word of 11 letters on the alphabet of English language (or any other European language based on the Latin letters).

Then
$$\varepsilon_{\mathbf{w}} \equiv (1,0,0,1,0,0,0,0,0,0,1).$$

Looking in fact at the structure of the word, we have:

$A = A, AB \neq RA, ABR \neq BRA, ABRA = ABRA, ABRAC \neq DABRA, ABRACA \neq$

$\neq ADABRA, \ldots, ABRACADABR \neq BRACADABRA, ABRACADABRA =$

$= ABRACADABRA.$

From this example we easily understand that the leading number $\varepsilon_{\mathbf{w}}$ of a word $\mathbf{w}$ must have some precise properties. In particular, whatever word $\mathbf{w}$ of length k is considered, one must have $\varepsilon_{\mathbf{w}}(k) = 1$, since the sub-word made with the *first* k letters and the sub-word made with the *last* k letters are the same object, namely they obviously coincide with the entire word $\mathbf{w}$. Furthermore, we have $\varepsilon_{\mathbf{w}}(1) = 1$ or $\varepsilon_{\mathbf{w}}(1) = 0$ depending on the fact that the first letter w_1 coincides, or not coincide, with the last letter w_k.

One can also understand that an arbitrary vector ε, with k binary coordinates and such that $\varepsilon(k) = 1$, does not necessarily coincide with the leading number $\varepsilon_{\mathbf{w}}$ of a word $\mathbf{w}$, of k letters. However, if $\varepsilon = \varepsilon_{\mathbf{w}}$ for some word $\mathbf{w}$, we can find also some other word $\mathbf{w}'$, such that $\varepsilon = \varepsilon_{\mathbf{w}'}$. For example, the two words *AABB* and *AAAB* admit the same leading number, namely $(0,0,0,1)$.

If $\mathbf{w}$ and $\mathbf{w}'$ are such that $\varepsilon_{\mathbf{w}} = \varepsilon_{\mathbf{w}'}$, what have they in common? Of course they, in particular, should share the same length. We will discover in the next section a further remarkable aspect that they have in common. In order to prepare such a discussion, we consider the natural number $M_{\mathbf{w}}$, defined by

$$M_{\mathbf{w}} = \sum_{u=1}^{k} \varepsilon_{\mathbf{w}}(u) N^u, \tag{1}$$

where N is the cardinality of the Alphabet.

For example, for the word $\mathbf{w} = ABRACADABRA$, seen as a word on the English alphabet ($N = 26$), we have

$$M_{\mathbf{w}} = 26 + 26^4 + 26^{11}.$$

Thus we associate an integer number to any word $\mathbf{w}$. More precisely, we associate an integer number to any pair $(\mathbf{w}, N)$.

This operation may appear just as one related to word-games or, at a different level, to Cabalistic tradition. However, as we shall see next, the number $M_{\mathbf{w}}$ will be of use in the field of applied probability and its meaning can be understood just through very basic probability concepts.

2 Random sequences of letters and waiting times of word occurrences

In this section we deal with sequences of random letters from an alphabet. On this purpose, a few elementary concepts from Probability theory are needed. For those readers, who are not familiar with basic language and notation of this theory, it can be useful to read the short Appendix at the end of this article.

Let an alphabet $\mathscr{A} \equiv \{a_1,...,a_N\}$ and a word $\mathbf{w} \equiv w_1 w_2 ... w_k$ on $\mathscr{A}$ be given (the choice of $\mathscr{A}$ and $\mathbf{w}$ will remain fixed once for ever along this section).

At different instants, letters are progressively drawn at random from $\mathscr{A}$. At time $t = 1$ we get the (random) letter ω_1, at time $t = 2$ we get the (random) letter ω_2, and so on.

Here we admit that $\omega_1, \omega_2, ...$ are *independent* and *uniformly distributed*; i.e. we assume that, at any instant t, each letter $a_1,...,a_N$ has the same probability $\frac{1}{N}$ to be drawn, independently of what happened at the previous instants $1, 2, ..., t - 1$; more formally we can write

$$P(\omega_t = a_j) = \frac{1}{N}, \text{ for } t = 1, 2, ..., j = 1, 2, ..., N;$$

and, for any $m > 1$ and $(a_{j_1}, a_{j_2}, ..., a_{j_m})$,

$$P((\omega_1 = a_{j_1}) \cap (\omega_2 = a_{j_2}) \cap ... \cap (\omega_m = a_{j_m})) = \frac{1}{N^m}.$$

We consider now the number of drawings needed until we see a series of k consecutive letters forming the word $\mathbf{w}$. This number is a random quantity, that we denote by $T_{\mathbf{w}}$. More formally it is defined by:

$$T_{\mathbf{w}} \equiv \inf\{n \geq k | \omega_{n-k+1} = w_1, ..., \omega_n = w_k\}.$$

Example 2. Let us fix the word $\mathbf{w} = AAB$ and sample the letters from the alphabet $\{A, B\}$. Suppose that we observed the sequence

$$ABAAABAABBAA...$$

In this case we have then $T_{\mathbf{w}} = 6$.

It is obvious that $T_{\mathbf{w}}$ is greater or equal to k, and that it is a *random variable*. The probability distribution of $T_{\mathbf{w}}$ is then given by the probabilities:

$$P(T_{\mathbf{w}} = k), \qquad P(T_{\mathbf{w}} = k+1), ...$$

This distribution depends on N and on k. In particular, it is immediate to see that

$$P(T_{\mathbf{w}} = k) = \frac{1}{N^k}.$$

What may appear surprising, at least at a first glance, is that this probability distribution not only depends on N and on k, but for fixed N and k, it also depends on the particular structure of **w**: it is very simple to realize that already the probability $P(T_{\mathbf{w}} = k+1)$ does depend on the structure of **w**. An important fact is that the probability distribution of $T_{\mathbf{w}}$ is determined by the leading number $\varepsilon_{\mathbf{w}}$ (see, in particular, the paper [4]).

More specifically it can be proved that the expected value of $T_{\mathbf{w}}$, i.e.

$$\mathbb{E}\left(T_{\mathbf{w}}\right) \equiv \sum_{h=k}^{\infty} hP\left(X = h\right),$$

is exactly given by the number $M_{\mathbf{w}}$, as defined by the formula (1) given in the previous section! In other words, here is a clear meaning for $M_{\mathbf{w}}$: one can prove that

$$\mathbb{E}\left(T_{\mathbf{w}}\right) = \sum_{u=1}^{k} \varepsilon\left(u\right) \cdot N^{u} = M_{\mathbf{w}}. \tag{2}$$

A direct computation of $\mathbb{E}\left(T_{\mathbf{w}}\right)$ is not at all easy, however. The proof of the identity (2) has been given in the frame of Martingale theory, by exploiting a heuristic idea based on the notion of fair game. Such an idea is particularly smart and suggestive. It is well known by now to experts in Probability theory; in the following section, we will give a sketch of it, by looking at the special example of the word *ABRACADABRA* and after briefly recalling the concept of fair game.

3 Leading numbers and Fair Games

The analysis of situations where we exchange a random quantity with a deterministic one (or viceversa) is at the basis of applications of Probability such as Finance, Insurance, and Gambling and it is at the core of Risk Theory. The concept of *fair game* is central.

It is clear that, in these fields, the quantities of concern have the meaning of amounts of money; but one can still think of other types of problems where quantities of interest can be the amounts of other goods.

Let X be a random income that Ann can buy from Joe at the (deterministic) price m. The table of what is exchanged is as follows:

	+	−
Ann	*X*	*m*
Joe	*m*	*X*

We call this exchange a *game*. Ann gains the (possibly negative) amount $X - m$, whereas Joe gains the opposite amount $-X + m$. One says that the game is *fair* when $\mathbb{E}(X) = m$; the game is favorable for Ann when $\mathbb{E}(X) > m$.

The concept of fair game has a basic importance in the analysis of risk problems. Typically, however, the games that we encounter in common life are not at all favorable for us; think for instance of the cases when we buy the ticket of a lottery or stipulate a contract with an insurance company. An interesting property of such a concept is that a sum of independent fair games results in a fair game. More properties are formalized and studied in the frame of the *Martingale theory* (see e.g. [6] and references therein).

In order to explain the heuristic idea behind the formula (2), we can consider a special fair game.

Here, in order to be a winner, a player must enter the game at a right moment in connection with the time when the word $\mathbf{w} = ABRACADABRA$ occurs for the first time in a random sequence of letters. More precisely, we can describe the game as follows.

The random sequence of letters $\omega_1, \omega_2, \ldots$ is drawn at random from the 26 letters of English alphabet. At any time $s = 1, 2, \ldots$, a new player I_s enters the game and pays one euro. Then the letter ω_s is drawn. The game stops at the random time $T_{\mathbf{w}}$ when $\mathbf{w}$ occurs for the first time in the sequence.

At time $s = 1$, the player I_1 enters the game and pays one euro to the bank. Actually he bets on the event $(\omega_1 = A)$. If really $(\omega_1 = A)$ occurs, I_1 gains 26 euros, remains in the game, and bets the 26 euros on the event $(\omega_2 = B)$; otherwise he has to leave the game and looses the initial stake of one euro. If $\omega_1 = A$ and $\omega_2 = B$, I_1 gains 26^2 euros, bets all this amount of money on the event $(\omega_3 = R)$ and so on.

In the case when I_1 is extremely lucky, $T_{\mathbf{w}} = 11$, i.e.

$$\omega_1 \omega_2 \ldots \omega_{11} = ABRACADABRA,$$

then I_1 gains 26^{11} euros. Otherwise I_1 is forced to leave the game and even looses the initial stake of one euro.

The probability of observing the event $(T_{\mathbf{w}} = 11)$ is $\frac{1}{26^{11}}$. Then, we can say, I_1 pays one euro and receives the random amount X_1, where the probability distribution of X_1 is described by

$$P(X_1 = 26^{11}) = \frac{1}{26^{11}}, P(X_1 = 0) = 1 - \frac{1}{26^{11}}.$$

Thus $\mathbb{E}(X_1) = 1$ and the game is fair for I_1. Similarly it happens for the other players $I_2, I_3, \ldots$: each of them plays a fair game against the bank.

Let us now analyze what happens, from the collective view-point, to the bank at time $T_{\mathbf{w}}$, i.e. at the end of the game. We can summarize by saying that the bank pays a deterministic amount (at a random instant) and receives a random reward. More precisely the reward is exactly $T_{\mathbf{w}}$ euros, since $T_{\mathbf{w}}$ players - each paying one euro - have participated in the game. On the other hand the bank pays $26^{11} + 26^4 + 26^1$ euros. These three quantities are respectively due to the players $I_{T_{\mathbf{w}}-10}, I_{T_{\mathbf{w}}-3}, I_{T_{\mathbf{w}}}$.

In fact I_{T_w-10} is the completely lucky player who sees the occurrence of the whole word *ABRACADABRA* and wins 26^{11} euros. The other two players, I_{T_w-3} and I_{T_w}, who see the occurrence of the words *ABRA* and *A* respectively, win 26^4 and 26^1 euros.

Notice at this point that, among all the 11 players present in the game at time T_w, the winners are indicated by the positions of the 1's in the leading number ε_w!

Now, concerning the computation of $\mathbb{E}(T_w)$, one can argue that the overall game is fair for the bank. This put us in a position to conclude:

$$\mathbb{E}(T_w) = 26^{11} + 26^4 + 26^1.$$

This heuristic argument can be made completely rigorous by using the tools of Martingale theory (see [3,6]).

4 Apparent paradoxes: some concluding remarks

In the analysis of occurrences of words the theme of waiting times, that we have briefly sketched so far, is of interest for a number of different respects. It is by now a sort of classical topic of applied probability that had already been considered in the early times of this field; see for instance the discussion about *runs of successes* in coin-tossing [2]. Deep contributions, exploring unexpected features of this theme, continue to appear from time to time in the literature (see e.g. [5] and references therein).

A further issue of interest can be found in the multiplicity of fields where related results can be applied. The connections with several apparently distant concepts of Mathematics (such as martingales and fair games, algebraic properties and leading numbers) also constitute a further source of attraction.

Concerning the probability calculus related to occurrences of words, we can mention that different problems have been studied and solved. Here we only dealt with the computation of $\mathbb{E}(T_w)$, but also other different issues, apparently difficult, turned out to admit mathematically interesting and nice solutions.

Let us in particular fix, on the same alphabet, two different words w_1 and w_2 of a same length k. One can consider the game where w_1 and w_2 are respectively chosen by two different players, each one betting one euro. The winner of the game is the one whose word occurs first, so that the probability that the player 1 wins is given by

$$P(T_{w_1} < T_{w_2}), \tag{3}$$

... and the winner takes it all.

Thus the game is fair if and only if $P(T_{w_1} < T_{w_2}) = 1/2$. The game is favorable to player 1 if and only if $P(T_{w_1} < T_{w_2}) > 1/2$.

An interesting algorithm for the computation of (3) is presented in [1].

What we especially want to point out here, however, is the circumstance that several different facts, that may appear as surprising, conflicting, and controversial, emerge within the area of word occurrences. Only a few facts will be reported below, even though several other nice surprises can be found in the relevant literature. An interesting issue is, in particular, the *comparison* between the waiting times $T_{\mathbf{w}_1}$, $T_{\mathbf{w}_2}$ for two given words $\mathbf{w}_1$ and $\mathbf{w}_2$, still of a same length k and on a same alphabet of N letters.

Whereas all words of a same length k, as already noticed above, admit the same probability $\frac{1}{N^k}$ to appear at once at the first k drawing of letters, $T_{\mathbf{w}_1}$ and $T_{\mathbf{w}_2}$ can have different probability distributions, depending on the structures described by the leading numbers $\varepsilon_{\mathbf{w}_1}$, $\varepsilon_{\mathbf{w}_2}$ (see [4]).

One can be interested in detecting which are the words that - in the random sequence of letters - appear in a sufficiently small time or, at least, before than some others words.

On the purpose of such a study, one can hinge on different types of *comparisons* between random variables. In particular the following, reasonable, ones have been considered in the relevant literature.

a) $T_{\mathbf{w}_1} \leq_a T_{\mathbf{w}_2}$ if and only if $P(T_{\mathbf{w}_1} > T_{\mathbf{w}_2}) \leq P(T_{\mathbf{w}_1} < T_{\mathbf{w}_2})$;
b) $T_{\mathbf{w}_1} \leq_b T_{\mathbf{w}_2}$ if and only if $\mathbb{E}(T_{\mathbf{w}_1}) \leq \mathbb{E}(T_{\mathbf{w}_2})$;
c) $T_{\mathbf{w}_1} \leq_c T_{\mathbf{w}_2}$ if and only if, for each positive t, $P(T_{\mathbf{w}_1} > t) \leq P(T_{\mathbf{w}_2} > t)$.

The last definition corresponds to what is called the *stochastic ordering* among random variables.

Generally, for two random times T_1 and T_2, it is well known that $T_1 \leq_c T_2 \Rightarrow T_1 \leq_b T_2$ and $T_1 \leq_b T_2 \not\Rightarrow T_1 \leq_c T_2$. A different picture is encountered however in the special frame of waiting times of words. As, at least, to our knowledge no pair of words (of a same length) have been found such that

$$T_{\mathbf{w}_1} \leq_b T_{\mathbf{w}_2}, \; T_{\mathbf{w}_1} \not\leq_c T_{\mathbf{w}_2}$$

and some work in progress at present time may lead to consider the conjecture that $T_{\mathbf{w}_1} \leq_b T_{\mathbf{w}_2} \iff T_{\mathbf{w}_1} \leq_c T_{\mathbf{w}_2}$.

Heuristically, we can say that all the three relations $T_{\mathbf{w}_1} \leq_a T_{\mathbf{w}_2}$, $T_{\mathbf{w}_1} \leq_b T_{\mathbf{w}_2}$ and $T_{\mathbf{w}_1} \leq_c T_{\mathbf{w}_2}$ convey the information that, at least in a probability sense, $T_{\mathbf{w}_1}$ is smaller than $T_{\mathbf{w}_2}$. It can therefore appear as surprising that one can easily find examples showing that

$$T_{\mathbf{w}_1} \leq_c T_{\mathbf{w}_2} \not\Rightarrow T_{\mathbf{w}_1} \leq_a T_{\mathbf{w}_2},$$

$$T_{\mathbf{w}_1} \leq_a T_{\mathbf{w}_2} \not\Rightarrow T_{\mathbf{w}_1} \leq_b T_{\mathbf{w}_2},$$

and then, (since $\leq_c \Rightarrow \leq_b$), also $T_{\mathbf{w}_1} \leq_b T_{\mathbf{w}_2} \not\Rightarrow T_{\mathbf{w}_1} \leq_a T_{\mathbf{w}_2}$, $T_{\mathbf{w}_1} \leq_a T_{\mathbf{w}_2} \not\Rightarrow T_{\mathbf{w}_1} \leq_c T_{\mathbf{w}_2}$.

We furthermore point out that, while the relations $\leq_b$ and $\leq_c$ obviously satisfy the transitive property, such a property is not respected by the relation $\leq_a$. In fact one can find an integer n and choose a finite sequence of words $\mathbf{w}_1, \mathbf{w}_2, \ldots, \mathbf{w}_n$ such that $T_{\mathbf{w}_i} \leq_a T_{\mathbf{w}_{i+1}}, T_{\mathbf{w}_{i+1}} \not\leq_a T_{\mathbf{w}_i}$, for $i = 1, \ldots, n-1$, and moreover $T_{\mathbf{w}_n} \leq_a T_{\mathbf{w}_1}, T_{\mathbf{w}_1} \not\leq_a T_{\mathbf{w}_n}$ (see the arguments presented in [1]).

Reasoning on these arguments on the basis of simple examples, for instance about runs of successes in coin-tossing or of sequences of red numbers at a roulette table in a Casino, may allow the reader to appreciate the apparently paradoxical character of the above statements.

At this point one can wonder: why the above-mentioned mathematical facts may be perceived as surprising and contradicting common intuition? A possible response, in our opinion, can be the following one. Apart from being, or not being, familiar with probability theory, common intuition of people about (discrete) waiting times is biased by the experience with geometrically distributed random variables. We saw above that, for the case of waiting times of words occurrences, the properties a), b), c) are not at all equivalent; actually they can be even conflicting. In the geometric case, on the contrary, corresponding properties a'), b'), c') turn out to be equivalent, as shown by Proposition 1 in the Appendix.

It is rather common that a detailed and logically rigorous analysis of intriguing problems leads sometimes to surprising and apparently paradoxical conclusions. Cases of this type are frequent in the frame of applied probability and statistics, where the analysis is made complicate by the subtle effects of uncertainty. In this respect, we conclude by claiming that ruses and surprising results can be specially met in the frame of word occurrences.

5 Appendix: a few basic notions about discrete random variables

Given a *random event* E, $P(E)$ denotes the probability that E happens (or that it results in a *success*); $1 - P(E)$ is then the probability that E does not happen. The *indicator* of E, denoted by $\mathbf{1}_E$, is a *binary random variable*, i.e. a random variable whose possible values are the only two values 0 and 1: $\mathbf{1}_E = 1$, if E results in a success, and $\mathbf{1}_E = 0$ otherwise. Thus we can also write

$$P(E) = P(\mathbf{1}_E = 1).$$

Let us consider a sequence of *events* $E_1, E_2, \dots$. It is said that $E_1, E_2, \dots$ are *Bernoulli trials* when they are *equiprobable* and *independent*. This means that, for a given p $(0 < p < 1)$, one has

$$P(E_1) = P(E_2) = \dots = p$$

and, for any choice of m,

$$P(\mathbf{1}_{E_1} = 1, \dots, \mathbf{1}_{E_m} = 1) = p^m.$$

Let X be a *discrete* random variable whose possible values are $x_1, x_2, \dots$. The *probability distribution* of X is specified by the probabilities

$$P(X = x_1), P(X = x_2), \dots$$

The *expected value* of X is given by

$$\mathbb{E}(X) \equiv \sum_{h=1}^{\infty} x_h \cdot P(X = x_h),$$

provided that the sum of the series is absolutely convergent. Basic properties of the operator $\mathbb{E}(\cdot)$ are the following: in the special case when $X = \mathbf{1}_E$ we have

$$\mathbb{E}(X) = P(E).$$

In the special case when X is a deterministic (or, better, a *degenerate)* random variable, i.e. when there exists $x \in R$, such that $P(X = x) = 1$, then

$$\mathbb{E}(X) = x.$$

The expected value $\mathbb{E}(\cdot)$ is a *linear operator*: if a, b are two real numbers and X, Y two random variables, then for the random variable $Z = aX + bY$ we have

$$\mathbb{E}(Z) = a\mathbb{E}(X) + b\mathbb{E}(Y).$$

A random variable X, whose possible values are $1, 2, \ldots$, is said to have *a geometric distribution with parameter* θ, when

$$P(X = k) = \theta (1 - \theta)^{k-1}, k = 1, 2, \ldots$$

In this case, it is easily seen that

$$\mathbb{E}(X) = \frac{1}{\theta}.$$

Let $E_1, E_2, \ldots$ be a sequence of random events and denote by T the number of trials until the first success among $E_1, E_2, \ldots$ occurs; more formally: $T = r$ if and only if we observe

$$\mathbf{1}_{E_1} = 0, \mathbf{1}_{E_2} = 0, \ldots, \mathbf{1}_{E_{r-1}} = 0, \mathbf{1}_{E_r} = 1.$$

The random variable T is then the *waiting time* until the first success.

In the case when $E_1, E_2, \ldots$ is a sequence of Bernoulli trials of probability p, then the probability distribution of T is *geometric with parameter* $\theta = p$.

We fix our attention on an experiment that can result in d different outcomes $A_1, \ldots, A_d$ with probabilities $p_1, \ldots, p_d$, respectively ($p_1 + \ldots + p_d = 1$). Suppose that the experiment can be repeated an arbitrary number of times, maintaining the same probabilities of outcomes and ensuring a condition of independence among the different trials. Consider now, for the indices $v = 1, 2$, the sequence of events

$$E_1^{(v)}, E_2^{(v)}, E_3^{(v)}, \ldots \tag{4}$$

where

$$E_j^{(v)} \equiv (\text{the experiment results in the outcome } v \text{ at the } j\text{-th trial}).$$

Thus, for $v = 1, 2, E_1^{(v)}, E_2^{(v)}, E_3^{(v)}, \ldots$ is a sequence of Bernoulli trials, of probability p_v. Furthermore, with an obvious meaning of notation, we have that the random waiting time $T^{(v)}$,

$$T^{(v)} = \inf\{n \geq 1 | \mathbf{1}_{E_n^{(v)}} = 1\},$$

has a geometric distribution and

$$\mathbb{E}\left(T^{(v)}\right) = \frac{1}{p_v}.$$

Fix now attention on the comparison between $T^{(1)}$, $T^{(2)}$. We notice that, by definition, the event $\left(T^{(1)} = T^{(2)}\right)$ is impossible. In this respect one can easily prove

$$P\left(T^{(1)} < T^{(2)}\right) = \frac{p_1}{p_1 + p_2}.$$

Concerning the above sequence of Bernoulli trials, we can then state a very simple result:

Proposition 1. *The following conditions are equivalent:*

$a')P\left(T^{(1)} > T^{(2)}\right) \leq P\left(T^{(2)} > T^{(1)}\right);$
$b')\mathbb{E}\left(T^{(1)}\right) \leq \mathbb{E}\left(T^{(2)}\right);$
$c')$ *for any positive* t, $P(T^{(1)} > t) \leq P(T^{(2)} > t);$
$d')p_2 = P(E_j^{(2)}) \leq P(E_j^{(1)}) = p_1.$

References

1. R. Chen, A. Zame, On fair coin-tossing games. J. Multivariate Anal. **9**(1):150–156, 1979.
2. W. Feller, An introduction to probability theory and its applications. Vol. I. Third edition. John Wiley & Sons Inc., New York, 1968.
3. S. Yen, R. Li, A martingale approach to the study of occurrence of sequence patterns in repeated experiments. Ann. Probab. **8**(6):1171–1176, 1980.
4. S. Robin, J. J. Daudin, Exact distribution of word occurrences in a random sequence of letters. J. Appl. Probab. **36**(1):179–193, 1999.
5. V. T. Stefanov, S. Robin, S. Schbath, Waiting times for clumps of patterns and for structured motifs in random sequences. Discrete Appl. Math. **155**(6-7):868–880, 2007.
6. D. Williams, Probability with martingales. Cambridge University Press, Cambridge, 1991.

E Pluribus Unum

Marco Li Calzi

1 Introduction

E pluribus unum is Latin for "Out of many, one". This sentence is best known as one of three shown on the Seal of the United States, that was adopted by an Act of Congress in 1782. It appears on the obverse side of the seal, as well as on official documents and U.S. currency. Considered since long the unofficial motto of the United States (whereas the official motto since 1956 is "In God we trust"), it was originally conceived to represent the idea that a single nation would emerge out of many states.

The symbolism in the seal is reinforced by a recurring motif that honors the original thirteen States in the Union with thirteen stars in the "glory" above the eagle's head, thirteen stripes on the shield, and thirteen arrows in the eagle's talon; moreover, custom has added thirteen leaves and olives on the olive branch. Numerologists will take pride in noting that the motto itself consists of thirteen letters, although this seems coincidental.

Fig. 1 The obverse side of the Seal of the United States

Marco Li Calzi
Department of Management, Università Ca' Foscari Venezia (Italy).

Emmer M. (Ed.): Imagine Math. Between Culture and Mathematics
DOI 10.1007/978-88-470-2427-4_18, © Springer-Verlag Italia 2012

The main theme of this article is the exploration of situations where a myriad of interactions between different people leads to the emergence of a (possibly unexpected) aggregate behavior. The best known example is the metaphor of the *invisible hand* first coined by the economist Adam Smith (1723–1790) to describe the social mechanism of the market, where the individual preferences are composed and the needs of the society are met even if each single agent is only attending his own business. However, as economists know well, such harmony is difficult to attain in practice. When dealing with human affairs, we should perhaps focus on a more neutral expression. The title of this article is my best contribution in this respect.

I would be remiss in failing to disclose that there is at least one contender that trumps the outcome of my efforts. This favorite of mine is the brilliant title of a book by Thomas C. Schelling (born 1921). He is a professor of foreign affairs who was awarded the 2005 Nobel Memorial Prize in Economic Sciences (shared with Robert Aumann) for "having enhanced our understanding of conflict and cooperation through game-theory analysis." His crowning achievement in this domain is the monograph *The Strategy of Conflict*, published in 1960, where he introduced key concepts in the analysis of conflict such as focal point and credible commitment; see [Mye09]. But the title I envy is *Micromotives and Macrobehavior*, appeared in 1978, where by a stroke of genius he let many individual heterogenous motivations adjoin the aggregate behavior of the system.

A simple example may help to illustrate how a multitude of local rules coalesce into a single aggregate behavior, sometimes in unexpected ways [Bon02]. There is a score of people in a closed space, such as a gymnasium. Consider two slightly different variants of a simple game. The first game has each person threatened by an attacker and protected by a defender: this agent moves around trying to keep the defender between the attacker and himself. The second game has each defender trying to interpose herself between the victim and the attacker. For clarity, we may dub "seek protection" the first version and "provide protection" the second version.

If we let people interact, what kind of aggregate behavior shall we expect to observe? As it turns out, "seek protection" tends to keep agents spread out and running around quite a lot; while "provide protection" gather a pretty tight crowd where people move very little. An elegant visualization is accessible at http://www.icosystem.com/demos/thegame.htm. (If a picture is worth a thousand words, a short movie must count as a million.) The point of the example is that, in spite of the remarkable similarity among the rules of the two games, they generate a completely different macrobehavior.

Examples of emergent macrobehavior abound. Closer to home (unless you live in Venice) is the case of traffic behavior. Each of the single drivers is pursuing a private objective relying on his own personal heuristics. At the macro level we observe a variety of behaviors, ranging from a smooth flow to inextricable traffic jams. For more examples, we can appropriate the blurb in the poster advertising *Unraveling Complex Systems* as the theme of the *2011 Mathematics Awareness Month*:

> We are surrounded by complex systems. Familiar examples include power grids, transportation systems, financial markets, the Internet, and structures underlying everything from the environment to the cells in our bodies. Mathematics and statistics can guide us in un-

derstanding these systems, enhancing their reliability, and improving their performance. Mathematical models can help uncover common principles that underlie the spontaneous organization, called emergent behavior, of flocks of birds, schools of fish, self-assembling materials, social networks, and other systems made up of interacting agents.

The Mathematics Awareness Month is sponsored each year by the Joint Policy Board for Mathematics, a collaborative effort of the American Mathematical Society, the American Statistical Association, the Mathematical Association of America, and the Society for Industrial and Applied Mathematics; see www.mathaware.org.

A final word of caution is in order. Philosophers talk about *explanandum* and *explanans*. The first is the phenomenon that needs to be explained, and the second is the explanation of that phenomenon. The study of complex systems is particularly good at generating emergent behaviors (*explanandum*), but somewhat less effective in providing the *explanans*. You may spot similar limitations in this article.

2 Traffic management

How do we get traffic queues? An obvious answer is that sometimes there are simply too many vehicles with respect to the capacity of the existing road network. Another recurring suspect are traffic lights, that may engender queues when their timing is not attuned to the traffic flow. Accidents, or other catastrophic events, are a third likely cause.

Some queues, however, originate in highways in the absence of any of the factors above. Imagine a smooth flow of cars moving at constant average speed over a highway. Each car travels just a few meters away from the preceding one. When, by some accidental event, the vehicle in the n-th position suddenly slows down, it forces the subsequent car $n + 1$ to do the same. The deceleration is quite more rapid than the acceleration, so it takes some time for each vehicle to get back to the average speed.

Looking from an imaginary helicopter hovering above the traffic, we see the gap between $n - 1$ and n growing wider, because car $n - 1$ keeps its speed while car n is braking up. Meanwhile, car $n + 1$ is forced to slow down to avoid collision with n so the gap between these two vehicles closes, and we see $n + 1$ queuing behind n. By induction, this generates a traffic wave moving upstream. The car in front of the wave can accelerate and move forward, while the subsequent vehicles must await until the space in front is cleared. The queue starts traveling backwards. Depending on the parameters, this may generate surprisingly long queues. Assume that car n has braked because a butterfly disturbed its driver, and the "butterfly effect" takes a new meaning. For a visual animation, we recommend http://www.traffic-simulation.de, where one can also explore similar queueing phenomena due to incoming traffic from a ramp, or to an uphill road that slows down trucks, or to closing one lane in a two-lane highway.

A rather different phenomenon is known as *Braess' Paradox*. We illustrate it with a simple example borrowed from [EK10]. Consider the road network in Fig. 2,

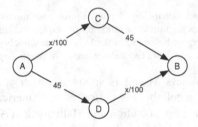

Fig. 2 A simple road network

where each arc is labeled with the travel time (in minutes) that it takes to go from one end to the other when there are x cars using it. There are 4000 cars that start at A and must reach B. Each driver simultaneously chooses whether to drive through C or D, in a conscious attempt to minimize his traveling time. We say that the traffic is *at an equilibrium* when, given the distribution of vehicles over the available routes, no driver can unilaterally switch to a different path and save time.

For instance, if all drivers choose to go through C, there are $x = 4000$ cars on the arcs A-C and C-B. Each car needs 40 minutes to go from A to C and a further 45' from C to B, so the total driving time for a vehicle is 95'. Facing this situation, any single driver would rather switch to the other (empty) route where the total traveling time for one car is $45 + (1/100) = 45.01$ minutes. Hence, everybody on the same route is not an equilibrium. As symmetry suggests, the only equilibrium for this network is that cars distribute equally over the two routes. When there are 2000 vehicles on the A-C-B route and another 2000 on A-D-B, the driving time for each car is $45 + 2000/100 = 65'$; a unilateral switch to the other route would give a higher traveling time of $45 + 100/101 = 45.01$. Thus, given the traffic network in Fig. 2, we expect people to learn to split over the two routes so that each driver takes 65' to complete his route.

Suppose that some well-meaning politician, bent on improving the miserable commuting times of the local community, decides to build a new highway between C and D. To keep things simple (without loss of generality), we assume that the new highway is one-way from C to D and that it is so large and capacious that the time to cover it is zero regardless of the number of vehicles using it. We can represent the new situation by drawing a new edge from C to D and label it zero in Fig. 2. The new network now offers three ways to go from A to B: the two old routes (A-C-B and A-D-B) and the new one (A-C-D-B).

Table 1 compares a few possible configurations. For instance, the second line looks at the situation where 500 cars take the old route through C, another 500 go through the old route through D, and the remaining 3000 drive through the new highway. The last two columns report that the driving time over the old routes is $3500/100 + 45 = 80'$, while the new route takes only $3500/100 + 3500/100 = 70'$. As it turns out, in each of the examples shown (and, indeed, in any possible configuration), going through the highway is always faster than using any of the old routes.

Table 1 Driving times over different traffic configurations

Traffic on route			Driving time	
through C	*through D*	*through CD*	*through C or D*	*through CD*
0	**0**	**4000**	**85**	**80**
500	500	3000	80	70
1000	1000	2000	75	60
1500	1500	1000	70	50
2000	2000	0	65	40

This is not surprising: the whole purpose of opening the new road is to lower the driving times.

The paradox shows up when we consider the consequences of this uniform superiority. Regardless of the traffic configuration, taking the highway is always better than trying any of the old routes. Hence, any driver ends up using the highway and all 4000 cars will be traveling on the same route. As shown in the first line of Table 1, this implies that in equilibrium the driving time is 80' for everybody. Yet, in the good old times (without the highway), the unique equilibrium had a traveling time of 65' — a whole 15' less! Paradoxically, opening up the highway makes everybody worse off. The emergence of this paradox in a general network is reasonably likely [SZ83]; see also Wikipedia for a few real-life examples. A recent result is that, if the driving time over any arc is a linear function of the number of cars using it, the worst-case scenario for a network upgrade is to have the equilibrium driving time going up by at most 33% [RT02].

The morale is that some actions may have unintended consequences if we ignore the collateral effects on individual behavior. Building the highway upgrades the infrastructure. Considered in isolation, this is an improvement. However, when the rest of the network is left unchanged, this local upgrade attracts so many drivers that it ends up clogging the old routes A-C and D-B. These routes easily suffer from congestion and thus the performance of the system degrades. Such collateral effects are called *externalities* by economists. They are wonderfully illustrated by a cartoon (ubiquitous on blogs lamenting traffic jams) where each of the drivers stuck in a huge traffic jam is thinking aloud: *If these idiots would just take the bus, I could be home by now.*

Our last foray in traffic management is borrowed from *Micromotives and Macrobehavior*:

> Standing in line at a ski lift — a long line — I overheard somebody complain that the chairs ought to go faster. It would take a bigger engine, but at those fees the management could afford one. The complaint deserves sympathy but the proposal doesn't work: speeding the lift makes the lines longer.   (p. 67)

How so? Let us work out an example. There are 20 skiers in the field. It takes one minute to get in and out of the lift, ten minutes to go up, and five to get down. Ignoring for simplicity other activities such as sipping coffee in the adjacent hut, skiers must be engaged in one of these three activities, or otherwise waiting in line. Since each of the 20 skiers needs one minute to get in and out of the lift, a full cycle that brings all skiers up lasts 20 minutes. During this time, a skier spends 1' to get in and out, 10' to go up, 5' to get down, and the remaining $20 - 16 = 4$ minutes standing in line. The queue itself has on average 4 people.

Let us see what happens if the management brings in new engines and the time to go up is halved. It still takes 20 minutes to complete a full cycle and get all skiers on board. But now a skier spends 1' to get in and out, 5' to go up, 5' to get down, and the remaining $20 - 11 = 9$ minutes standing in line. The new engines roar aloud and do their job, but there are now nine people standing in line and grumbling. Where do the extra five skiers waiting in line come from? Before the engines were installed, they used to be on the lift moving at a slower speed and enjoying the view (instead of suffering the ignominy of queueing). The reader may check that the paradox of faster lifts leading to longer queues arises when the number of skiers in the field is $n \geq 11$.

3 Family issues

Wedding customs vary greatly, but a piece of the western tradition seems to me a little odd. While two people are joined in marriage at the front, families and friends usually stand (or sit) behind them in two separate groups. This form of segregation during the ceremony is often imposed by seating arrangements. It may be explained in many ways, often amenable to the symbolism of wedding as an alliance between two families.

However, somewhat mysteriously, a similar segregation takes place *spontaneously* when the wedding reception is organized as a standing event where people can flow freely. Dreaded by many brides, this segregation occurs in spite of their best efforts to have people from the groom's and bride's families mix. People gather together in small groups and recombine, but it is often the case that most groups see a prevalence of one side. Why?

An important piece of the explanation is that even a tiny preference for one's neighbors to be from one's own family may lead to segregation. It is agents' preference to congregate with their relatives that tends to keep the two groups apart. This simple argument was suggested in 1969 by Schelling [Sch69] in a context inspired by racial dynamics. His argument was developed using a physical model with coins placed on graph paper, where dimes and pennies were moved around simulating people's inclinations. This approach brilliantly predates the rich computer simulations nowadays abundant in the social sciences, usually known as agent-based models. Let us illustrate Schelling's argument.

There are two kinds of agents, labelled X and O, who represent the bride's and the groom's relatives. They can move around in the reception hall, that we visualize as a grid. Each agent takes place in a cell of the grid and interacts with his immediate

Table 2 A configuration of agents in a grid

X	X				
X	O_1		O		
X	X	O	O	O	
X	O			X	X
	O	O	X_2	X	X
		O	O	O_3	

X	X	O	O	X	X
X	X	O	O	X	X
O	O	X	X	O	O
O	O	X	X	O	O
X	X	O	O	X	X
X	X	O	O	X	X

neighbors. For instance, the left-hand side of Table 2 depicts a grid with $6 \times 6 = 36$ cells and 11 agents for each family.

The immediate neighbors of an agent are those in the eight cells around him (or less, if he is adjacent to the walls). We assume that an agent feels comfortable when his immediate neighborhood contains at least 30% people from his family. Anyone who has experienced the discomfort of being the "odd man out" in a small group entertaining a conversation should be able to relate to this assumption.

If the current arrangement makes him uncomfortable, an agent moves to another cell in the grid. For instance, starting from the situation depicted on the left of Table 2, only three agents (O_1, X_2, O_3) are uncomfortable and will move to a different position. Doing so, they create a new configuration where perhaps other agents feel uncomfortable and in turn decide to move. The dynamics may takes time to unfold, but in the end agents' choices lead to a stable configuration. This represents the underlying macrobehavior emerging from agents' micromotives.

Note that agents are not prejudiced against the other group: they need not be part of a majority to feel comfortable. With eight neighbors, having three of one's own relatives around is enough to stay on. Such individual preferences are compatible with perfect integration, as shown on the right-hand side of Table 2. However, when guests are randomly distributed, a different macrobehavior emerges.

This is illustrated in Fig. 3, where agents from the two groups are colored light and dark grey, respectively. They smile if they feel comfortable and frown otherwise. The initial configuration on the left is mixed up. The dynamics eventually leads to a stable configuration such as that one on the right, where a clear amount of segregation has taken place. The figures are created using the simulation tool available at http://www.rensecorten.dds.nl. This is in turn based on NetLogo, a multi-agent programmable modeling environment widely used by students, teachers and researchers worldwide. Freely accessible at http://ccl.northwestern.edu/netlogo, it offers a library with a large variety of simulative examples, ranging from biology to social sciences.

Our next example is drawn from [ML75] and takes place after marriage has been consummated. It illustrates the effect of introducing a bit of variability in individual behaviors. Suppose that all parental couples have an innate preference for males, and that they keep bearing children until their progeny has more boys than girls. The baseline case assumes that each couple has the same probability $p = 1/2$ of giving birth to a male or a female. Since each family stops growing when boys are

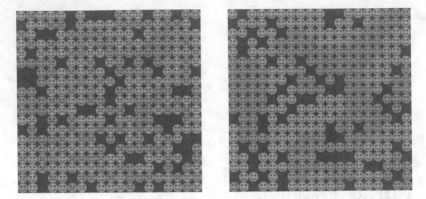

Fig. 3 Initial (left) and final (right) configurations

a majority, it is not surprising that this leads to a situation where males are more numerous than females.

Consider now the more plausible assumption that each couple k has a different probability p_k of giving birth to a boy, although on average the value of p_k is still $1/2$. For instance, if a couple has probability $(1/2 - \varepsilon)$, there is another one for whom the probability is $(1/2 + \varepsilon)$. Anything else is unchanged: each couple keeps procreating until the majority of its progeny is male. The following paradoxical effect emerges: while each family is actively seeking a majority of boys, the society as a whole ends up with a majority of girls!

How is it possible? The exact argument requires some familiarity with random walks, but here is an intuitive explanation. Families with a propensity to generate boys $(p_k > 1/2)$ tend to reach their target of a male majority quickly, and thus stop procreating soon. They generate more boys than girls, but the absolute numbers are small. On the other hand, families with a propensity to generate females keep chasing their target of a male majority and thus end up procreating a lot more girls. The tiny majority of boys generated by the male-biased families is rapidly overwhelmed by the large numbers of girls born in female-biased families. (My favorite simulation of this model using NetLogo is available at http://www.agsm.edu.au/bobm/teaching.)

4 Odds and ends

Our last section parades a few vignettes describing a wide range of applications amenable to the study of emergent behavior in the social sciences; see [Eps07] for more examples.

The Culture Model [Axe97] is a seminal study on how beliefs (or attitudes) in a population shift over time, getting closer or diverging. It has been used to explain how opinions get spatially clustered, the emergence of bandwagon effects (fashion fads), and the spontaneous division of a culture into sub-cultures. A recent application discusses how knowledgeable people can use individuals with wide social networks to disseminate information quickly and effectively, in a conscious attempt to induce "social epidemics" such as new political aggregations or outbursts of moral outrage on the internet. A curious spinoff of this line of research has looked into modeling standing ovations, when people from an audience spontaneously stand to acknowledge an outstanding performance [MP04].

A well-known, albeit controversial, foray into archeology looks at the rise and fall of the Anasazi civilization in the southwestern United States. Until its disappearance around AD 1350, the Anasazi society widely fluctuated in population and settlement size. The study combines quantitative information on environmental fluctuations with plausible behavior rules for Anasazi households and computes a detailed historical "trajectory" that matches the known facts.

Other models purport to enucleate a few key driving elements in political action. An elegant example is a study of civil violence [Eps07] considering two models. The first one illustrates how a subjugated population seemingly coordinates its rebellion against a central authority; see Rebellion in the NetLogo library for a visual animation. The reason why a repressive regime stays in place is not that his police is stronger than the people, but that the former can coordinate much better than the latter. Fiddling with its parameters, one is reminded of Tocqueville's dictum: "It is not always when things are going from bad to worse that revolutions break out. On the contrary, it oftener happens that when a people which has put up with an oppressive rule over a long period, it takes up arms against it." (*The Ancien Régime and the French Revolution.*) The second model describes the dynamics of inter-group violence: the analogies with recent historical examples (Hutu vs. Tutsi, or Serbs vs. Bosniaks) where local ethnic cleansing led to genocide are impressive.

A promising application for the study of emergent behavior is panic control, where one attempts to design solutions that reduce the risk associated with orderly crowds suddenly switching behavior due to panic [H00]. (For a recent example of poor design, recall the Love Parade stampede in July 2010, where 21 people lost their lives and more than 500 were injured.) Space prevents us from a longer discussion, but more information including visual animations is available at http://angel.elte.hu/ panic/.

We close with a puzzle lifted from Schelling's *Micromotives and Macrobehaviors*. At a conference, it is often the case that seats in the first few rows are empty. Can you figure out why?

References

[Axe97]   R. Axelrod, The dissemination of culture: A model with local convergence and global polarization. J Conflict Resolut **41**, 203–226, 1997.

[Axt02]   R.L Axtell et al., Population growth and collapse in a multiagent model of the Kayenta Anasazi in Long House Valley. PNAS **99**, 7275–7279, 2002.

[Bon02]   E. Bonabeau, Predicting the unpredictable. Harvard Bus Rev **80**, 1–9, 2002.

[EK10]   D. Easley, J. Kleinberg, Networks, Crowds, and Markets: Reasoning about a Highly Connected World. Cambridge University Press, Cambridge, 2010.

[Eps07]   J.M. Epstein, Generative Social Science: Studies in Agent-Based Modeling. Princeton University Press, Princeton, 2007.

[ESP02]   J.M. Epstein, J.D. Steinbruner, M.T. Parker, Modeling Civil Violence: An Agent-Based Computational Approach. PNAS **99**, 7243–7250, 2002.

[H00]   D. Helbing, I. Farkas, T. Vicsek, Simulating dynamical features of escape panic. Nature, 407, 487–490, 2000.

[ML75]   J.G. March, C.A. Lave, Introduction to Models in the Social Sciences. Harper Collins, New York, 1975.

[MP04]   J. Miller, S. Page, The standing ovation problem. Complexity, 9, 8–16, 2004.

[Mye09]   R.B. Myerson, Learning from Schelling's *Strategy of Conflict*. J Ec Lit **47**, 1109–1125 2009.

[RT02]   T. Roughgarden, É. Tardos, How bad is selfish routing?. J ACM **49**, 236–259, 2002.

[Sch69]   T.C. Schelling, Models of segregation. Amer Econ Rev **59**, 488–493, 1969.

[SZ83]   R. Steinberg, W.I. Zangwill, The prevalence of Braess' paradox. Transp Sci **17**, 301–318, 1983.

Aperiodic Tiling

Gian Marco Todesco

There is an aesthetic pleasure when contemplating orderly structures that contain some disorder. A completely disordered pattern is typically not very interesting, but neither is a very regular one, like a check board. The check board and most images that we will meet in the following are examples of *tessellations*. A plane tessellation (or *tiling*) is a covering without gaps or overlaps, by figures called *tiles*. Tessellations can have very different degrees of order and disorder and illustrate well the concept expressed in the first statement.

1 Periodic tiling

A tessellation is *periodic* if it has translational symmetry, i.e. if it remains unchanged after certain translations. A periodic tessellation is a highly ordered pattern that can be summarized by a finite region that tessellates the whole plane just by translations. Periodic tessellations can also present non-translational symmetries. All these symmetries interact with each other in a complex way, creating constraints. For instance, only a two, three, four and six-fold rotational symmetry is allowed: a periodic tiling cannot have a five-fold rotational symmetry.

The subject is mathematically interesting and has many applications in architecture, decorative arts, technology and crystallography. Periodic tessellations appear everywhere in our lives. In fact, they are so common that one might not realize that there exist also non-periodic tessellations.

Gian Marco Todesco
Digital Video S.r.l., Rome (Italy).

Emmer M. (Ed.): Imagine Math. Between Culture and Mathematics
DOI 10.1007/978-88-470-2427-4_19, © Springer-Verlag Italia 2012

2 Non-periodic tiling

There are many classes of non-periodic tessellations. The tilings below present scale symmetry instead of translational symmetry (Fig. 1).

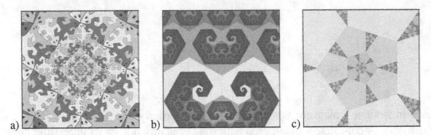

Fig. 1 Self-similar non-periodic tessellations: a) image inspired by "Smaller and Smaller" by M.C. Escher; b) non-archimedean tiling discovered by L. Saffaro; c) a plane tessellation with regular pentagons

The following examples exclude scale transformation, but are still non-periodic (Fig. 2).

The tiles of the tessellations above can be rearranged to form periodic patterns. A tile set that formed only non-periodic tiling would be called an *aperiodic* set. One could think that such a set does not exist and indeed this conjecture was formulated in 1961 by the mathematician Hao Wang. Wang was working on a decision problem in symbolic logic and found an interesting connection between his problem and some particular tiles called *Wang dominoes*. In 1966 Robert Berger demonstrated that the conjecture is false and found a set of 20426 Wang dominoes that can tessellate the plane only non-periodically. The number of tiles has then been reduced to 104 by Berger himself, to 92 by Donald Knuth, to 40 by Hans Läuchli, and to 13 by Karel Culik. In 1971 Raphael M. Robinson found a new set of six aperiodic tiles, that are not Wang dominoes [2].

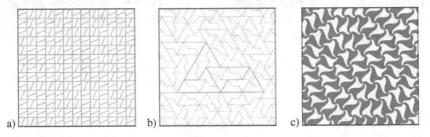

Fig. 2 Non-periodic tilings with identical tiles: a) a trapezoid; b) the sphinx, a self-replicating pentagon; c) a curved triangular shape forming a spiral pattern created by H. Voderberg

Some years later Roger Penrose tried to find a tile set including the pentagon (a polygon with the five-fold symmetry forbidden in the periodic tessellation) [3]. He was inspired by a tessellation with pentagons in the *Harmonices mundi* (1619) by Kepler. It is interesting to note that also Albrecht Dürer in his *Painter's Manual* (1525) describes a partial tiling with pentagons. Eventually, in 1973 and 1974, Penrose discovered three new periodic sets, two of which with only two tiles.

Three years later Robert Ammann discovered several new sets with similar properties.

The quest for an even simpler tile set has continued up to very recently. A single tile that form an aperiodic set is jokingly called an *einstein* (the name does not refer to the famous physicist, but to the German words *ein Stein* that mean "one stone", thus one tile). In 2010, Joshua Socolar and Joan Taylor claimed to have solved the einstein problem.

The discovery of aperiodic tessellation has been a breakthrough. One of the Penrose aperiodic sets is the subject of this conference.

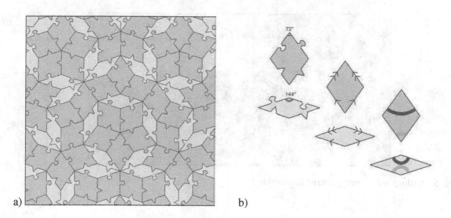

a)   b)

Fig. 3 a) The Penrose rhombs tiling; b) the thick and thin rhombs. Three different ways to enforce the matching rules

3 Thick and Thin Rhombs

The tessellation above is one of Penrose's aperiodic tessellations and is made of the thick and thin rhombs displayed on the right. Either of them or both in some combinations could tessellate the plane periodically, thus, to form an aperiodic set, the two tiles must be altered in some way. For instance one can add tabs and blanks to the edges as in the jigsaw puzzle pieces. A more common and pleasing approach (used in the following images) is to add colored curves that must match in color and position to the tiles.

These changes define a set of *matching rules* that allow only some configurations. The rules admit only seven (instead of 54) combinations of pieces around a vertex, shown in the image below.

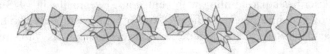

Fig. 4 The eight available vertex configurations

The tessellations made with these rhombs have many interesting features.

First of all there is an uncountably infinite number of different tessellations and each tessellation, being non-periodic, contains infinite different patterns. Many patterns are meaningful and aesthetically pleasant.

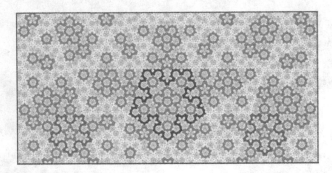

Fig. 5 A tiling with some patterns highlighted

Then, different patterns and different tilings look very similar to each other. Indeed it is possible to demonstrate a very surprising theorem: any limited region (even very large ones) is replicated infinite times in any tessellation. Such order within disorder is known as *quasiperiodicity*.

The patterns have a local five-fold rotational symmetry. In other words there are regions that remain unchanged after a 72° rotation. Some of the tilings present even a global five-fold rotational symmetry (i.e. the 72° rotation does not change the whole tessellation).

To summarize, these tessellations have a strong and complex internal structure, that is not easy to grasp. Even assembling the tiling is not an easy task. A wrong choice can lead to patch that cannot be further extended and we discover this only after many moves. Assembling a given number of tiles is very much like finding a way out in a maze where the dead ends can be arbitrarily long.

There is a phenomenon, called *inflation* and *deflation*, that can be used to assemble large patterns without errors. Inflation and deflation transform a tiling into a new one at a different scale. The rules for the deflation are represented in the image below.

The tiles are replaced by a larger number of smaller tiles forming a new tessellation. This process can be iterated an arbitrary number of times, generating a sequence of tilings. It is possible to start with a single tile and create a tiling that covers an arbitrarily large region.

The process implies a hierarchical structure: each tile in a tessellation can be considered the "parent" of some of the tiles in the deflated tessellation. (Note that a tile has two parents). This hierarchical structure plays the role that the translation has for periodic patterns.

Fig. 6 The deflation rules

4 Applications

The difficulty of assembling many pieces and the beauty of the patterns they form lead naturally to use them for puzzles. Indeed Penrose licensed a company called Pentaplex to produce and sell jigsaw puzzles based on his tilings. Some years later, Kleenex designed a toilet paper tissue using rhombs tiling (which apparently saves paper and makes it fluffier without bunching when rolled up). Penrose, who had patented his tiles (in spite of the criticism raised for patenting a mathematical concept), sued the company and eventually won. The patents have now expired.

Non-periodic tilings have been used as decorative patterns for floors and walls. Some examples are the Storey Hall at the Federation Square and the Royal Melbourne Institute of Technology (Australia), the Liberal Arts and Science Building in Education City in Doha (Qatar) and the Ravensbourne College of Design and Communication in London.

There are also surprising examples from the past. Some decorations in medieval Islamic architecture, and in particular patterns in the Darb-e Imam shrine, built in 1453, at Isfahan, Iran, present five-fold symmetry and a hierarchical structure (large tiles are decorated with pattern formed by smaller identical tiles), strongly resembling to Penrose tilings (formalized five centuries later) [4].

The most important application started ten years after the Penrose's discovery, solving a big crystallographic problem and eventually leading to a paradigm shift and the Nobel Prize in chemistry in 2011.

In 1982 Dan Shechtman and his colleagues Ilan Blech, Denis Gratias and John Cahn observed ten-fold electron diffraction patterns (forbidden by the usual crystallographic restrictions) in a rapidly solidified metal alloy [5]. Two years later Ilan Blech, using computer simulations, suggested that the diffraction patterns could result from an aperiodic structure. The scientific community took some time to accept the concept, but eventually, in 1992, the International Union of Crystallography altered its definition of crystal, broadening it to include *quasicrystals*, i.e. ordered but not periodic arrangements of atoms.

Since then hundreds of quasicrystals with various compositions and symmetries have been discovered, opening the door to potential applications (with side-effects in everyday life: e.g. they make excellent non-scratch coating for frying pans). Recent evidence found that they can even form naturally under suitable geological conditions. Today the study of quasicrystals is a very active and promising field and aperiodic tiles are no more a mathematical curiosity but an object of intense study.

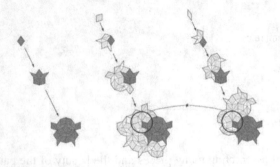

Fig. 7 Deflation can be used to extend the tiling in different ways

5 Assembling the tiles

The image above shows how repeated applications of the deflation process can tessellate a finite region starting from a single ancestor tile. This technique can generate an arbitrarily large tessellation, but if we want to extend a given tessellation outside its boundaries then we need to go back to the ancestor and re-parent it. In other words we need to find a new tile that, when deflated, generates the old ancestor and other siblings. The grand-children of these siblings extend our tessellation.

To extend the pattern we must know the ancestor, therefore we need a global knowledge of the pattern. Moreover the process is not deterministic. There are two possible new ancestors that generate the old one. Each one can properly extend the initial tiling, but the two extensions differ. To cover the infinite plane we have to make an infinite number of choices.

The deflation process does not generate a given tiling in a foreseeable way: at each step the whole tiling is still undetermined because of the infinite sequence of choices that are still to be made.

6 The cut-and-project method

In 1981 the Dutch mathematician Nicolaas Govert de Bruijn discovered that there is a relation between some aperiodic patterns and regular lattices at higher dimensions. This result sheds a different light on the subject and also offers a new method to generate incrementally the patterns using only local information.

The following aperiodic tessellation offers a first glance of the technique.

It looks like a pile of cubes arranged along a steep slope and actually it can be generated this way, projecting onto a properly inclined plane the faces of a regular cubic grid that interpolate the plane itself.

To generate the Penrose rhombs, with their typical 5-fold rotational symmetry, we need five axes instead of three. Therefore we must consider a cubic grid in a space with 5 dimensions: a challenge for our imagination. Before applying the method in five dimensions, we must acquaint ourselves with it in a much simpler configuration.

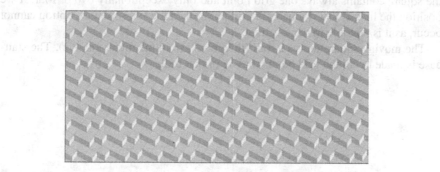

Fig. 8 A non-periodic tiling that can be seen as parallel projection of a set of cubes

7 The Fibonacci tiling

The Fibonacci Tiling is a one-dimensional non-periodic tiling of the line, made of an infinite number of adjacent segments. Only two types of segments with different lengths are used and the segments sequence does not repeat.

To generate this tiling with the cut-and-project method, we must consider a two-dimensional square lattice, i.e. a grid of evenly spaced points. Only horizontal or vertical straight lines can be drawn by connecting adjacent points of the grid. A diagonal line can only be approximated by a staircase-like polygonal chain. The staircase is made of two kinds of segments: vertical and horizontal, both of unit length. These segments, projected on the line, form a one-dimensional tessellation of the line with two tiles.

In the following, we suppose that the line slope is less than $45°$. On this line, the projection of the vertical staircase steps is shorter than the projection of horizontal steps. Let us call L and S (for long and short) the horizontal and vertical steps respective projections. They are the tiles.

The line slope determines the sizes of the two tiles, but also their relative frequencies. Indeed the slope of the line is the ratio between the frequencies of S and L. If the tessellation is periodic this ratio must be a rational number because in each (finite) repeating pattern there is the same given number of S and L. If the line has an irrational slope, then the tessellation must be non-periodic.

In the above picture the slope of the line is the inverse of the golden ratio, the well-known irrational number Φ.

Having the line and the grid we must build the staircase made of vertices and edges that belong to the grid. We can select the vertices first and then select the edges that connect them. Thus the core of the cut-and-project method is a criterion to select grid vertices.

A suitable criterion is based on a unit square whose center slides along the line. The square edge length is equal to the distance of adjacent points in the grid, so the square contains always one grid point and only exceptionally two or four. If we position the line so that it touches a grid point, then the four-point exception cannot occur, as it is possible to demonstrate.

The moving square leaves a trail (the gray strip in the picture above). The staircase is made by all the grid points that lie in that trail.

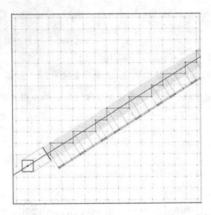

Fig. 9 The Fibonacci tiling: a non-periodic one-dimensional tiling of the line

8 One-dimensional tiling from a three-dimensional lattice

The Fibonacci tiling is an effective introduction to the cut-and-project method. It shows that an aperiodic pattern can be derived from a regular grid in a higher dimensional space and explains how the fascinating mixture of order and disorder of the aperiodic tessellations derives from the ordered lattice structure and from the irrational slope between the grid and the tiles.

In the general case we must consider three spaces: the n-dimensional space that contains the grid, the m-dimensional *direct space* that we want to tessellate and its orthogonal complement which is an $(m - n)$-dimensional space called the *internal space*. To select the points we use an n-dimensional grid unit cell, sliding along the direct space.

The internal space can be used to effectively select the grid points. The key concept is that the projection of the unit cell on the internal space remains still while the cell slides along the direct space. Thus, instead of matching the grid points against the unit cell trail (the gray strip of the previous example), we can project everything on the internal space and then select the grid points whose projections are contained in a *control shape* that is the projection of the unit cell.

Our target case has a five-dimensional grid, a bi-dimensional direct space and therefore it has a three-dimensional internal space. It is convenient to investigate first another simpler case which involves three dimensions only, but presents an internal space more complex than a line. We can consider a line tessellation generated by a three-dimensional lattice. In this case the unit cell is a cube and the internal space has two dimensions, i.e. it is a plane.

Following the same strategy of the previous example, we consider a three-dimensional cubic grid and a straight line whose slope is irrational with respect to all the three axes. The line is then approximated by a polygonal chain that connects adjacent points of the grid and plays the role of the staircase. Because of the irrational slope of the line, the polygonal chain will follow it and curl almost randomly around it.

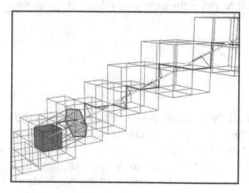

Fig. 10 A non periodic tiling of the line projected from a three-dimensional grid

The tessellation is generated by projecting the polygonal chain segments to the line. The projected segments have three different lengths that depend on the angle between the line and the three coordinate axes. Therefore the tessellation is made of three different tiles. As in the previous example, their sequence is not periodic.

The internal space is a plane perpendicular to the line and the control shape (i.e. the projection of the unit cube to the plane) is a hexagon. We select the grid points whose projections on the plane are contained in the hexagon.

We notice that the projections of the selected points fill evenly the hexagon. This is due to the irrationality of the line slope.

9 Bi-dimensional tiling from a five-dimensional lattice

Now we can face the thick and thin rhombs challenge. We must consider a five-dimensional lattice and a plane with proper slope with respect to the lattice. We choose the plane perpendicular to the space diagonal $(1,1,1,1,1)$ so to generate five-fold rotational symmetry on the plane. This constraint does not determine completely the plane orientation in a five dimensional space. We have two more degrees of freedom that we can use so that all the faces of the five-dimensional grid have the shape of the thick and thin rhombs, when projected onto the plane.

The internal space has three dimensions and it is perpendicular to the plane (this is not so easy to visualize). The control shape is the projection of the unit 5-cube on the internal space, analogous of the hexagon in the previous example. In this case it is a rhombic dodecahedron, i.e. a convex polyhedron delimited by twelve rhombs.

To select a grid point we have to project it on the internal space and check if it is contained in the dodecahedron. Then we select all the grid square faces whose vertices have all been selected. They form a faceted surface that is analogous to the Fibonacci tiling staircase. This faceted surface, projected on the plane, generates a thick and thin rhombs tessellation.

It is possible to demonstrate that, if the plane has the correct distance from the grid origin, then the generated tessellation respects the matching rules and therefore it is a Penrose tessellation.

10 "My God - it's full of stars!"

The projections on the internal space of all the selected points are contained in the rhombic dodecahedron. The actual distribution of the projections in the polyhedron is interesting and contains information about the tiling.

The projections do not fill all the space inside the control shape, but they distribute on four parallel planes. These planes contain the vertices of the polyhedron and their intersections with the polyhedron are pentagons. The projections of the selected points fill evenly the four pentagons.

Analyzing the tiling, we have already noticed that there are only seven different type of tile configuration around each vertex (see Fig. 4). Categorizing the vertices according to their tile configuration, there are seven types of vertices. If we consider again the internal space and we color the projected points according to the type of the correspondent tessellation vertex, the colors do not mix chaotically, but form a beautiful pattern made of five-pointed stars.

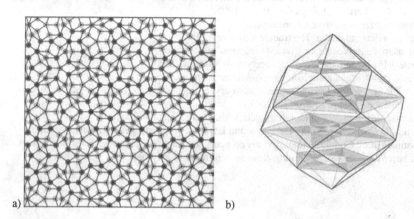

a)   b)

Fig. 11 a) Different vertex configurations painted with different colors; b) the related points in the internal space

References

1. L. Saffaro, Tassellature centrali e non-archimedee. Le Scienze 271, pp. 32-40, 1991.
2. B. Grünbaum, G.C. Shephard, Tilings and Pattern's. W.H. Freeman, New York, 1986.
3. R. Penrose, Pentaplexity. Eureka 39, 1978; reprinted in: The Mathematical Intelligencer 2(1), 1979; italian edition in: M. Emmer (ed.), L'occhio di Horus. Istituto Enciclopedia Italiana, Roma, 1989, 196-201. See also R. Penrose making the aperiodic tiling with kite and dard in the film "Symmetry and Tessellations", by M. Emmer, 1982.
4. P.J. Lu, P.J. Steinhardt, Decagonal and quasi-crystalline tilings in medieval Islamic architecture. Science 315, 1106–1110, 2007.
5. D. Shechtman, I. Blech, D. Gratias, J. Cahn, Metallic Phase with Longrange Orientational Order and no Traslational Symmetry. Phys. Rev. Lett. **53**, 1951, 1984.
6. M. Gardner, Penrose tiles to trapdoor cipher. W.H. Freeman, New York, 1989.
7. N.G. de Bruijn, Algebraic theory of Penrose's non-periodic tilings of the plane. I, II, Nederl. Akad. Wetensch. Indag. Math. **43**(1), 39–52, 53–66, 1981.
8. E. Harriss, D. Frettlöh, Tilings Encyclopedia. In rhombs, and wedges, and half-moons, and wings. http://tilings.math.uni-bielefeld.de/
9. Wikipedia, Aperiodic tiling, http://en.wikipedia.org/wiki/Aperiodic_tiling
10. Wikipedia, Penrose tiling, http://en.wikipedia.org/wiki/Penrose_tiling
11. E.O. Harriss, On Canonical Substitution Tilings. PhD thesis, Imperial College London, 2003. http://www.mathematicians.org.uk/eoh/files/Harriss_Thesis.pdf

12. L. Effinger-Dean, The Empire Problem in Penrose Tilings. Honors thesis, Williams College, 2006. http://www.cs.washington.edu/homes/effinger/files/empire.pdf
13. T. Smith, Penrose tilings and wang tilings. http://www.tony5m17h.net/pwtile.html
14. Math Pages. On de Bruijn Grids and Tilings. http://www.mathpages.com/home/kmath621/kmath621.htm
15. J. Sólyom, Fundamentals of the physics of solids. Springer, Heidelberg Berlin, 2007.
16. G. Egan, de Bruijn notes. 2008. http://gregegan.customer.netspace.net.au/APPLETS/12/deBruijnNotes.html
17. P. Bourke, Non Periodic Tiling of the Plane. 1995. http://paulbourke.net/texture_colour/nonperiodic/
18. C. Kaplan, +Plus magazine. The trouble with five. 2007. http://plus.maths.org/content/os/issue45/features/kaplan/index
19. A. Boyle, +Plus magazine. From quasicrystal to Kleenex. 2000. http://plus.maths.org/content/os/issue16/features/penrose/index
20. P.J. Lu, P.J. Steinhardt, Decagonal and Quasi-crystalline Tilings in Medieval Islamic Architecture. Science 315, pp. 1106-1110, 2007.
21. A. Feldman, My Bathroom Floor. http://alexfeldman.org/personal/floor.html
22. S. Collins, Bob - Penrose Tiling Generator and Explorer. http://stephencollins.net/penrose/
23. D. Austin, Feature Column. Monthly essay on mathematical topics, Penrose Tiles Talk Across Miles. http://www.ams.org/samplings/feature-column/fcarc-penrose

Mathematics and Medicine

Connecting Ventricular Assist Devices to the Aorta: a Numerical Model

Jean Bonnemain, Simone Deparis and Alfio Quarteroni

Mechanical circulatory support, in particular ventricular assist devices (VAD), has been recently proposed as an alternative to transplantation in the treatment of terminal heart failure in the context of the lack of donors and raising number of patients on the waiting list. Although these systems have proved their efficiency through a rigorous patient selection, the complication rate remains high and experience shows that many of them are related to haemodynamic modifications due to VAD implantation. Furthermore, VAD themselves have been widely studied, while the flow near the anastomosis VAD-aorta is still not well-known, although many complications arise at this site. We present here the mathematical settings and some preliminary results of a numerical model of the anastomosis between the outflow cannula of left ventricular assist devices (LVAD) and the aorta.

1 Introduction - The Clinical Problem

Heart failure is a clinical syndrome that expresses the inability of the heart to provide enough blood to the organs in order to satisfy their metabolic needs. In other terms, the pump is failing. The causes of heart failure are numerous (coronary artery disease and myocardial infarction, valvular disease, cardiomyopathy, myocarditis, to mention a few) as well as their symptoms (e.g. dyspnea, fatigue, peripheral edema) [8].

Within the population studied by the European Society of Cardiology (>900 million people in 51 countries), heart failure prevalence is around 2-3% and affects at least 15 millions patients [4]. In the USA, 2.5% of the adult population (5.7 million people) suffer from heart failure and 670'000 new cases occur each year. Further-

Jean Bonnemain
CMCS - MATHICSE - EPFL, Lausanne (Switzerland),
Cardiovascular Surgery Department, CHUV, Lausanne (Switzerland),
Simone Deparis
CMCS - MATHICSE - EPFL, Lausanne (Switzerland),
Alfio Quarteroni
CMCS - MATHICSE - EPFL, Lausanne (Switzerland),
MOX, Politecnico di Milano, Milano (Italy).

Emmer M. (Ed.): Imagine Math. Between Culture and Mathematics
DOI 10.1007/978-88-470-2427-4_20, © Springer-Verlag Italia 2012

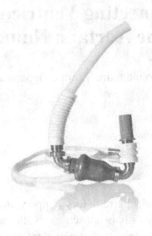

Fig. 1 The VAD HeartMate
II. Reprinted with the permission of Thoratec Corporation

more, prevalence is higher in the elderly - 6-10% of people older than 65 years. Among the deaths reported in 2004, 292'000 were caused directly or indirectly by heart failure. The estimated overall cost of heart failure in the USA in 2009 was 37.2 billion dollars [10].

The first line of the treatment of heart failure includes lifestyle modifications, like sodium and fluid restriction, smoking cessation or weight loss. These are necessary but usually not sufficient. The second line of treatment is the use of drugs, like angiotensin-converting enzyme (ACE) inhibitors, angiotensin-receptor blockers, diuretics and β-blockers. Finally, invasive treatment includes cardiac resynchronisation therapy, that consists in the implantation of a pacemaker (biventricular pacing, with or without implantable cardioverter defibrillator) and transplantation [14]. However, heart donors are scarce (2210 transplantations in 2007 in the USA) and, in June 2008, 2607 patients were still on the waiting list [10]. An alternative approach has been recently suggested in order to compensate the lack of donors and to provide a treatment for terminal heart failure: the *mechanical circulatory support*.

Various types of mechanical circulatory support exist with different clinical indications, but their goal is always the same: assist (or replace) the pump function of the heart. They are classified as follows:

- Intra-Aortic Balloon Pump (IABP);
- Extracorporeal Membrane Oxygenation (ECMO);
- Ventricular Assist Devices (VAD);
- Total Artificial Heart (TAH).

IABP and ECMO are short-term assist devices (hours to days) while VAD and TAH have an intermediate or long-term use (days to years). Depending on the indications, they can be considered as:

- *Bridge to transplantation*, for patients on waiting list;
- *Bridge to recovery*, as temporary circulatory support;

- *Bridge to decision*, gives time to the clinician to find the best therapeutic option;
- *Bridge to eligibility*, for patients initially ineligible for transplantation who become eligible after VAD implantation, due to the normalization of several clinical parameters;
- *Destination therapy*, for patients who are ineligible for transplantation [20].

In this paper, we will focus on Ventricular Assist Devices (VAD) (see Fig. 1). They can assist right, left or both sides of the heart, according to the underlying pathology. Furthermore, they can deliver pulsatile or non-pulsatile flows, and they can be paracorporeal, partially or totally implantable. It is important to notice that newer, smaller, continuous flow pumps (e.g. see Fig. 2) showed better results, as reduction of complications, durability and mortality [21].

All these systems have proved their efficiency through a rigorous patient selection. However, the complication rate remains high, e.g. [7], in particular they may take the following form:

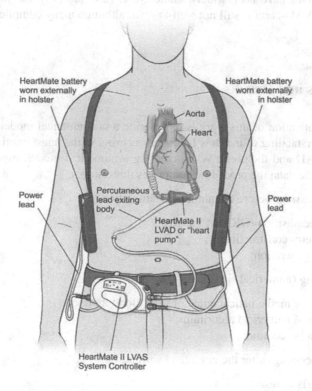

Fig. 2 Left Ventricle Assist Device, Reprinted with the permission of Thoratec Corporation. The inflow cannula is inserted in the apex of the deficient ventricle and the outflow cannula is anastomosed to the ascending aorta. Blood flow exits the ventricle through the inflow cannula, is actively pumped by the pumping chamber and goes through the outflow cannula to the aorta. A percutaneous drive line carries the electrical cable to the battery packs and electronic controls

- Bleeding (30%);
- Thromboembolism (3-35%);
- Infections (18-59%);
- Right ventricular failure (20-30%);
- Primary device failure (6% at 6 months to 64% at 2 years).

VAD as a *destination therapy* showed clinical benefits [20], however there is still room for better outcomes and for a reduction of the costs, which, for the time being, remain high [1].

As mentioned before, complications remain a major issue. Many of them are related to haemodynamic modifications due to the VAD implantation. In fact, abnormalities in flow and shear patterns can lead to platelet activation and to the formation of clots. In particular, regions of turbulent flow, recirculation and stagnation have a high thrombogenicity. Finally, subsequent shear stress distribution on the arterial wall can have short term negative effects (e.g. thrombus formation) and in the long-term has an impact on arterial remodeling and atherosclerosis [13]. While VAD themselves have been widely studied, e.g. [23] and [15], the flow near the anastomosis VAD-aorta is still not well-known, although many complications arise at this site.

2 Methods and Results

The main motivation of this work is to describe a mathematical model that allows a better understanding of the flow behavior occuring in the anastomotic region between the VAD and the aorta. When creating a numerical model, especially with patient-specific data, the procedure comprises three stages:

- *pre-processing* (before the numerical simulation)

  - data acquisition (DICOM images),
  - geometry construction,
  - mesh generation;

- *processing* (numerical simulation itself)

  - set-up of mathematical equations for blood motion,
  - set-up of numerical algorithms,
  - parallel execution;

- *post-processing* (after the numerical simulation)

  - analysis of results,
  - model validation.

A short description follows.

2.1 Pre-processing

For patients under ventricular assistance, the most frequent image acquisition method is *CT-scan* (computed tomography), because it is fast and vessels can be seen using intravenous contrast. CT-scan uses X-rays that provide a set of bidimensional images (slices) of a given region of the body (e.g. thorax, see Fig. 3).

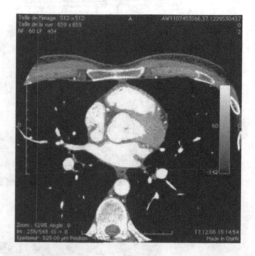

Fig. 3 Example of a CT-scan image at the level of the aortic valve

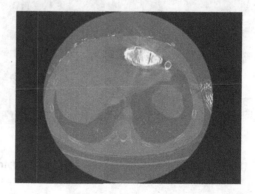

Fig. 4 CT-scan slice of a patient with a continuous flow LVAD at the level of the LVAD itself. On the top of the image, the VAD itself and the inflow cannula that connects the left ventricle to the LVAD. Note the noise and artifacts induced by the presence of the device

Unfortunately, the presence of the device, due to its metallic components, induces a lot of noise and artifacts on images (See Fig. 4). Consequently the geometry of the region of interest (outflow cannula of the LVAD, aorta and its branches) cannot be reconstructed directly on the native image (this would indeed produce an aberrant geometry). Therefore the images need to be cleaned by filters of various kind.

Treatment of images is performed by reducing the level window (i.e. focusing only on a defined grey range) and enhancing the contrast of the greyscale image using contrast-limited adaptive histogram equalization. In particular this filter allows to enhance the differences between the vessel with contrast (the region of interest) and the surrounding tissues (to be neglected in the segmentation). Then a gradient anisotropic diffusion filter is applied to reduce the noise, resulting in a smoothed image. Fig. 5 shows these different filters.

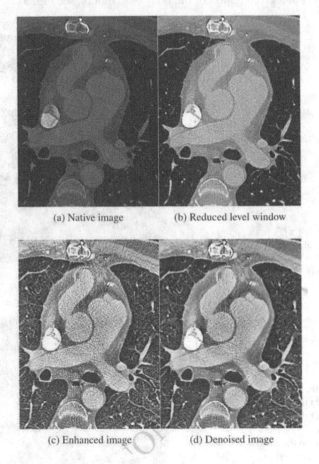

(a) Native image   (b) Reduced level window

(c) Enhanced image   (d) Denoised image

Fig. 5 Effects of the different filters applied to the DICOM images, seen at the level of the anastomosis between the outflow cannula of the LVAD and the aorta

Having convenient images for segmentation, the geometry of the region of interest can be reconstructed. The gradient of the previously treated images is then calculated and the watershed segmentation algorithm [22] is applied to the result. To improve the obtained segmentation, mathematical morphology filters are then performed and the final surface is extracted. All these steps are performed using the InsightToolKit library[1]. These methods allow to reconstruct only the geometry

---

[1] www.itk.org

of the fluid, i.e., the volume inside the artery corresponding to the blood (the lumen). The arterial wall itself is usually not seen on DICOM images like CT-scan or MRI, consequently its geometry cannot be directly reconstructed. Since we want to perform numerical simulations that takes into account arterial wall deformation (i.e. the so-called Fluid-Structure Interaction, FSI), we have to artificially reconstruct the arterial wall, assuming that arterial wall thickness is proportional to the local vessel radius [9].

Finally we used the library vmtk[2] to create meshes for the fluid (blood) and structure (arterial wall, outflow cannula) and to a-priori identify the boundary layers.

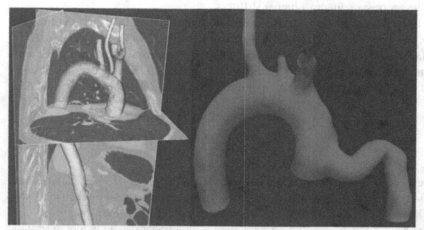

(a) Geometry of an healthy aorta obtained from a CT-scan exam.   (b) Geometry of the fluid domain (the lumen).

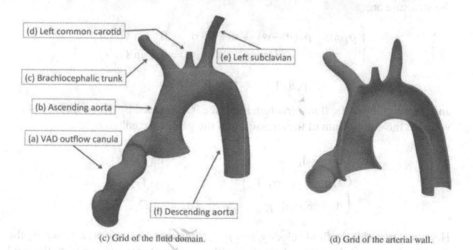

(c) Grid of the fluid domain.   (d) Grid of the arterial wall.

Fig. 6 Geometry and computational grids

2.2 Processing

We aim at modeling the flow in the whole computational domain in Fig. 6 (comprising (a) cannula, (b) ascending aorta, (c) brachiocephalic trunk, (d) left common carotid, (e) left subclavian, (f) descending aorta) by a fluid-structure coupled problem. More precisely, the blood flow in the lumen is described by the Navier-Stokes equations for a Newtonian fluid in the Arbitrary Lagrangian Eulerian (ALE) frame of reference [16], while the arterial wall is modeled by a 3D elastodynamic equation. The same model is adapted for the artifical cannula (which is made of woven polyester), the only difference in our model being represented by a larger Youg modulus, corresponding to a higher stiffness.

Let Ω_f and Ω_s be the reference domain for the fluid and the structure, respectively, and $\Gamma_{FSI} = \partial \Omega_f \cap \partial \Omega_s$ the fluid-structure interface. In the case at hand, $\Omega_f(t)$ is in fact the lumen of the whole domain in Fig. 6 (comprising the cannula, the aorta and its branches) whereas $\Omega_s(t)$ indicates the arterial and cannula walls. At every time $t > 0$, the current configuration of the fluid domain, $\Omega_f(t)$, is given by the ALE map

$$\mathscr{A}_t : \Omega_f \to \Omega_f(t)$$
$$\mathbf{x} \to \mathscr{A}_t(\mathbf{x}) = \mathbf{x} + \mathbf{d}_f(x),$$

where $\mathbf{d}_f$ is the displacement of the fluid domain, therefore $\Omega_f(t) = \mathscr{A}_t(\Omega_f, t)$. Practically $\mathbf{d}_f = Ext(\mathbf{d}_s|_{\Gamma_{FSI}})$, where $\mathbf{d}_s$ is the solid displacement and the extension $Ext(\cdot)$ is an harmonic lifting (or extension) operator from Γ_{FSI} to Ω_f (see [6]).

The Navier-Stokes equations are coupled with a linear elastic model describing the structure's behavior [6]. The partial differential equations modeling the fluid and the structure are:

$$\begin{cases} \rho_f \partial_t \mathbf{u} + \rho_f (\mathbf{u} - \mathbf{w}) \cdot \nabla \mathbf{u} - \nabla \cdot \sigma_f = 0 & \text{in } \Omega_f(t), \\ \nabla \cdot \mathbf{u} = 0 & \text{in } \Omega_f(t), \end{cases}$$

$$\rho_s \partial_{tt} \mathbf{d}_s - \nabla \cdot \Pi = 0 \quad \text{in } \Omega_s,$$

and the coupling at the fluid-structure interface is expressed by the continuity of the velocity, the equilibrium of the stresses, and the geometric adherence:

$$\begin{cases} \mathbf{u} = \partial_t \mathbf{d}_s & \text{on } \Gamma_{FSI}(t), \\ \Pi \mathbf{n}_s = -J_f \sigma_f (\mathbf{F}_f)^{-T} \mathbf{n}_f & \text{on } \Gamma_{FSI}, \\ \mathbf{d}_f = Ext(\mathbf{d}_s|_{\Gamma_{FSI}}), \quad \mathbf{w} = \partial_t \mathbf{d}_f & \text{in } \Omega_f. \end{cases}$$

Here: $\mathbf{u} = \mathbf{u}(\mathbf{x},t)$ is the fluid velocity, $p = p(\mathbf{x},t)$ the pressure, $\mathbf{w} = \mathbf{w}(\mathbf{x},t) = \partial_t \mathbf{d}_f$ the domain velocity; ∂_t the partial derivative with respect to the time; ρ_f and ρ_s the fluid and solid densities; $\sigma_f = \sigma_f(\mathbf{u}, p) = -p\mathbf{I} + \mu\varepsilon(\mathbf{u})$ the fluid Cauchy stress tensor, μ the fluid dynamic viscosity, $\varepsilon(\mathbf{u}) = (\nabla \mathbf{u} + \nabla \mathbf{u}^T)/2$ the strain rate tensor, and $\Pi =$

$\Pi(\mathbf{d_s})$ the first Piola-Kirchhoff stress tensor of the structure. Moreover $\mathbf{F}_f = \nabla \mathscr{A}_t$ is the fluid domain gradient of deformation, $J_f = \det \mathbf{F}_f$ the jacobian, $\mathbf{n}_f$ and $\mathbf{n}_s$ the outward unit normals to the fluid and solid domains. Additional condititions on the external boundary are necessary to close the equations.

Regarding the numerical approximation, we use here a geometry-convective explicit (GCE) time discretization of the 3-D FSI problem [3]. It means that both the convective field and the fluid computational domain are extrapolated from the previous time step. The other terms are treated with a first order backward Euler scheme. These equations are discretized in space by a $\mathbb{P}_1$-$\mathbb{P}_1$ Finite Element (FE) method stabilized by interior penalty. This implies that the Navier-Stokes equations are reduced to a linear problem at each time step. The equations for the structure are also linear, requiring no special treatment. Due to the explicit treatment of the geometry, the discrete coupled problem is linear. It can therefore be solved by a GMRES method preconditioned by overlapping algebraic Schwarz preconditioners based on an inexact block factorization of the system (see [3]).

The software used is based on LifeV (www.lifev.org), a parallel finite element library providing implementations of state-of-the-art mathematical and numerical methods. It has been used already in medical and industrial marks to simulate fluid structure interaction and mass transport processes [3]. The kind of simulations we are interested in are very heavy in term of computational costs, therefore supercomputers like the Blue Gene/P (IBM) or Cray XT or XE series are necessary [2].

2.3 Post-processing

Post-processing of the data is also an important issue since the solutions given by the simulations are heavy in terms of data size. Therefore critical parameters have to be carefully identified in order to provide a relevant and meaningful analysis of the results both from the mathematical and clinical points of view. Their description with respect to the model that is applied is made in chapter 3. We used ParaView[3] to perform the post-processing of the data. It allows to analyze results both in a qualitative (e.g. flow pattern at a given location) and quantitative manner (flowrate at a given vessel).

Finally the model has to be validated. We are currently performing *in vitro* validation. It is conducted by comparing values obtained with the *in vitro* model with the *in silico* one, based on the same geometries and input parameters, using the PIV (Particle Image Velocimetry) method.

---

[3] www.paraview.org

3 Results

We present here preliminary results of our numerical simulations. On a background of a network made of the coupling of 1D models for the systemic vasculature [5], [11], and [19], we overlay a 1D model for the cannula on the ascending aorta. This model has the advantage to have a low computational cost; simulating 6 heartbeats takes typically 8 hours using 8 processors (64 CPU hours). It also allows to evaluate the systemic effects of the LVAD. As a drawback it cannot evaluate local tridimensional features, e.g., the zones of flow recirculation or stagnation at the anastomosis site, nor the charge loss due to the anastomosis. Fig. 7 shows the flowrate and pressure at several critical arteries. Note that it is possible to evaluate pressure and flowrate at the main arteries.

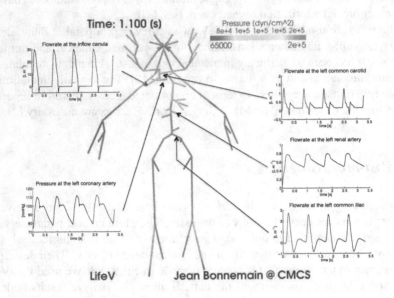

Fig. 7 Arterial tree model composed by 1D elements

We then enrich our model by using a 3D model of the region depicted in Fig. 6 comprising the connection of the outflow cannula, the aorta and its branches. Its coupling with a 1D model of the entire cardiovascular system (see Fig. 8) yields a geometric multiscale model [11], [12], [18], and [17]. The advantages of using such a multiscale model are numerous. In particular it is possible to evaluate the local effects of the insertion of the LVAD (with the 3D model) and evaluate its interaction with the cardiovascular system (with the coupling with the 1D model). The drawback is its elevated computational cost. E.g., using 64 cores on an IBM Intel Nehalem cluster composed by blades containing two quad-core 2.66 GHz nodes each, takes 72 hours to simulate 3 heartbeats (corresponding to as many as 4608 CPU hours). In particular, Fig. 10 shows secondary flows and stagnation zones in the region of anastomosis. In this case the aortic valve is closed and all the inflow comes from the cannula of the LVAD.

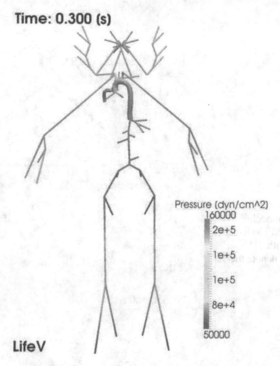

Fig. 8 Multiscale model. Arterial tree composed by a 3D FSI model for the domain of Fig. 6 and by 1D elements for the remaining circulation

4 Conclusions

This work represents the first step towards a clinical tool for patient-specific optimized VAD implantation. In that sense, we join the idea of the well-known expression: *from bench to bedside.*

More specifically in this study we have focused on the anastomosis of LVAD to the aorta. The techniques we developed allow us to run patient-specific simulations giving the opportunity to understand the behavior of the blood flow near the anastomosis and the interaction between LVAD and the cardiovascular system in a complete non-invasive way. In the long run, hopefully this will open the way to predictive surgery.

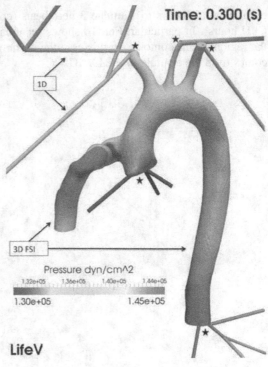

Fig. 9 Multiscale model. Focus on the 3D FSI model and its coupling with 1D elements. The stars denote the coupling interfaces

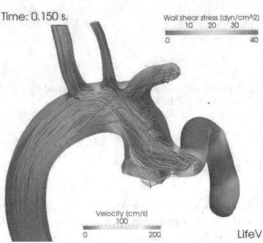

Fig. 10 Multiscale model. Focus on the 3D model showing the streamlines and the wall shear stress distribution (WSS), posterior view

Acknowledgements

We acknowledge Cristiano Malossi for having provided the multiscale framework and for his valuable help to set up the numerical simulations. We also thank Prof. Ludwig Karl von Segesser for his advice, especially in the clinical field. Finally we are grateful to Elena Faggiano and Paolo Crosetto for the help they provided, the Swiss National Fund grant 323630-133898, the European Research Council Advanced Grant Mathcard, Mathematical Modelling and Simulation of the Cardiovascular System, Project ERC-2008-AdG 227058, and the entire LifeV community.

References

1. A. Clegg, D. Scott, E. Loveman, J. Colquitt, P. Royle, J. Bryant, Clinical and cost-effectiveness of left ventricular assist devices as destination therapy for people with end-stage heart failure: a systematic review and economic evaluation. International journal of technology assessment in health care 23(2), 261–8, 2007.
2. P. Crosetto, S. Deparis, G. Fourestey, A. Quarteroni, Parallel algorithms for fluid-structure interaction problems in haemodynamics 33(4), 1598–1622, 2011.
3. P. Crosetto, P. Reymond, S. Deparis, D. Kontaxakis, N. Stergiopulos, A. Quarteroni, Fluid-structure interaction simulation of aortic blood flow. Computers and Fluids 43(1), 46–57, 2011.
4. K. Dickstein, A. Cohen-Solal, G. Filippatos, J.J.V. McMurray, P. Ponikowski, P.A. Poole-Wilson, A. Strömberg, D.J. van Veldhuisen, D. Atar, A.W. Hoes, A. Keren, A. Mebazaa, M. Nieminen, S.G. Priori, K. Swedberg, E.C. for Practice Guidelines, A. Vahanian, J. Camm, R.D. Cateriná, V. Dean, K. Dickstein, G. Filippatos, C. Funck-Brentano, I. Hellemans, S.D. Kristensen, K. McGregor, U. Sechtem, S. Silber, M. Tendera, P. Widimsky, J.L. Zamorano, D. Reviewers, M. Tendera, A. Auricchio, J. Bax, M. Böhm, U. Corrà, P.D. Bella, P.M. Elliott, F. Follath, M. Gheorghiade, Y. Hasin, A. Hernborg, T. Jaarsma, M. Komajda, R. Kornowski, M. Piepoli, B. Prendergast, L. Tavazzi, J.L. Vachiery, F.W.A. Verheugt, J.L. Zamorano, F. Zannad, Esc guidelines for the diagnosis and treatment of acute and chronic heart failure 2008. Eur Heart J 29(19), 2388–442, 2008.
5. L. Formaggia, D. Lamponi, A. Quarteroni, One-dimensional models for blood flow in arteries. Journal of Engineering Mathematics 47, 251–276, 2003.
6. L. Formaggia, A. Quarteroni, A. Veneziani (eds.), Cardiovascular Mathematics: Modeling and Simulation of the Circulatory System, Modeling, Simulation and Applications, vol. 1. Springer, Milan, 2009.
7. R. Gordon, B. Quagliarello, F. Lowy, Ventricular assist device-related infections. The Lancet Infectious Diseases 6(7), 426–437, 2006.
8. M. Jessup, S. Brozena, Heart failure. N Engl J Med 348(20), 2007–18, 2003.
9. G.J. Langewouters, Visco-elasticity of the human aorta in vitro in relation to pressure and age. Ph.D. thesis, Free University, Amsterdam, 1982.
10. D. Lloyd-Jones, R. Adams, M. Carnethon, G.D. Simone, T.B. Ferguson, K. Flegal, E. Ford, K. Furie, A. Go, K. Greenlund, N. Haase, S. Hailpern, M. Ho, V. Howard, B. Kissela, S. Kittner, D. Lackland, L. Lisabeth, A. Marelli, M. Mcdermott, J. Meigs, D. Mozaffarian, G. Nichol, C. O'donnell, V. Roger, W. Rosamond, R. Sacco, P. Sorlie, R. Stafford, J. Steinberger, T. Thom, S. Wasserthiel-Smoller, N. Wong, J. Wylie-Rosett, Y. Hong, Heart disease and stroke statistics–2009 update: A report from the american heart association statistics committee and stroke statistics subcommittee. Circulation 119(3), e21–e181, 2009.

11. A.C.I. Malossi, P.J. Blanco, S. Deparis, A two-level time step technique for the partitioned solution of one-dimensional arterial networks, 2011. Submitted
12. A.C.I. Malossi, P.J. Blanco, S. Deparis, A. Quarteroni, Algorithms for the partitioned solution of weakly coupled fluid models for cardiovascular flows. International Journal for Numerical Methods in Biomedical Engineering, 2011
13. K. May-Newman, B. Hillen, C. Sironda, W. Dembitsky, Effect of lvad outflow conduit insertion angle on flow through the native aorta. Journal of Medical Engineering & Technology 28(3), 105–9, 2004.
14. J. Mcmurray, M. Pfeffer, Heart failure. The Lancet 365(9474), 1877–1889, 2005.
15. R. Medvitz, J. Kreider, K. Manning, A. Fontaine, S. Deutsch, E. Paterson, Development and validation of a computational fluid dynamics methodology for simulation of pulsatile left ventricular assist devices. ASAIO Journal 53(2), 122–131, 2007.
16. F. Nobile, Numerical approximation of fluid-structure interaction problems with application to haemodynamics. Ph.D. thesis, École Polytechnique Fédérale de Lausanne, 2001.
17. T. Passerini, M. de Luca, L. Formaggia, A. Quarteroni, A. Veneziani, A 3D/1D geometrical multiscale model of cerebral vasculature. Journal of Engineering Mathematics 64(4), 319–330, 2009.
18. A. Quarteroni, A. Veneziani, Analysis of a geometrical multiscale model based on the coupling of PDE's and ODE's for blood flow simulations. Multiscale Modeling & Simulation 1(2), 173–195, 2003.
19. P. Reymond, F. Merenda, F. Perren, D. Rufenacht, N. Stergiopulos, Validation of a one-dimensional model of the systemic arterial tree. Am J Physiol Heart Circ Physiol 297(1), H208–222, 2009.
20. E.A. Rose, A.C. Gelijns, A.J. Moskowitz, D.F. Heitjan, L.W. Stevenson, W. Dembitsky, J.W. Long, D.D. Ascheim, A.R. Tierney, R.G Levitan, J.T. Watson, P. Meier, N.S. Ronan, P.A. Shapiro, R.M. Lazar, L.W. Miller, L. Gupta, O.H. Frazier, P. Desvigne-Nickens, M.C. Oz, V.L. Poirier, Randomized evaluation of Mechanical Assistance for the Treatment of Congestive Heart Failure (REMATCH) Study Group, R.E.: Long-term mechanical left ventricular assistance for end-stage heart failure. N Engl J Med 345(20), 1435–43, 2001.
21. M. Slaughter, J. Rogers, C. Milano, S. Russell, J., Conte, D. Feldman, B. Sun, A. Tatooles, Delgado III, R., Long, J., Wozniak, T., Ghumman, W., Farrar, D., Frazier, O.: Advanced heart failure treated with continuous-flow left ventricular assist device. The New England journal of medicine 361(23), 2241–2251, 2009.
22. L. Vincent, P. Soille, Watersheds in digital spaces: An efficient algorithm based on immersion simulations. IEEE Transactions on Pattern Analysis and Machine Intelligence 13, 583–598, 1991.
23. Y. Zhang, Z. Zhan, X. Gui, H. Sun, H. Zhang, Z. Zheng, J. Zhou, X. Zhu, G. Li, S. Hu, D. Jin, Design optimization of an axial blood pump with computational fluid dynamics. ASAIO J 54(2), 150–5, 2008.

Mathematics Elsewhere

Numeracy, Metrology and Mathematics in Mesopotamia: Social and Cultural Practices

Grégory Chambon

Introduction

Since the first decipherments of cuneiform writing in the XIX century, academics have been interested in the reconstruction of the ancient Near Eastern numerical and metrological systems as well as abstract ideas and techniques, which were identified, in modern times, as mathematics.

On the one hand, numerical and metrological data are usually assumed to directly reflect the concrete world that scribes have tried to describe, quantify and organise. Thus Assyriologists – historians working on the Ancient Near East – have often tried to approach ancient political and economical reality through studies of ancient weight- and measurement-systems; but, as they think that *mathematical* practices are by nature only for mathematicians, they are not interested in them.

On the other hand, since the great works carried out in the 1930s and 1940s[1], the aim of historians of mathematics has been to provide evidence for metrology and mathematics in scribal schooling. But for many decades, their attention has been almost exclusively focused on the internal features of mathematics, and then they have put aside historical and social concerns so as to compare the content and concepts of Mesopotamian and Greek mathematics. For example, in the secondary literature, the so-called *Babylonian mathematics* is often assumed to represent the first *true* mathematics, with abstract ideas and procedures, considered as *proto-Greek* because of the lack of an organised and independent discipline before the classical Greek period from 600 to 300 BC[2].

However, it is misleading for historians of mathematics to focus only on internal features and developments of mathematics with no exploration of the social, cultural and intellectual activities that have made possible its creation, transformation and diffusion; in other words, they need to contextualise mathematics. Conversely, Assyriologists have gained a deeper understanding of the ancient societies

Grégory Chambon
Université de Bretagne Occidentale, Brest (France).

[1] See [17], [30], [18].

[2] For this topic, see [28], 268-274.

Emmer M. (Ed.): Imagine Math. Between Culture and Mathematics
DOI 10.1007/978-88-470-2427-4_21, © Springer-Verlag Italia 2012

and cultures by considering mathematical techniques as culturally embedded practices; mathematical techniques are *expressions* of social and intellectual groups of a given culture[3].

In order to highlight the importance of such an interdisciplinary and contextualising approach, this paper explores some interactions between mathematical techniques, material culture and scribal activities through a close examination of two case-studies.

1 Mathematicians, historians and "Babylonian Mathematics"

1.1 The received picture of "Babylonian Mathematics" in the modern world

In the secondary literature, the term *Babylonian mathematics* usually refers to any mathematics developed by the people of Mesopotamia over the period of time extending from the beginning of the third millennium to the fall of Babylon, in 539 BC.

At first glance this terminology appears to be misleading. How can one use the simple word *Babylonian* to refer to a broad and complex three-thousand-year tradition at the crossroads of many cultures and languages of the Ancient Near East? In fact, Babylon, even during its existence, was not always the greatest city in Mesopotamia. Even if the city played an important role in developing and transmitting cultural, intellectual and religious knowledge, it was not at the crossroads of intellectual life and scholarship at any time of the Mesopotamian history[4].

For a long time Mesopotamia (from the Ancient Greek "between rivers") has referred to the area of the Euphrates and Tigris river system set in modern-day Iraq and north-eastern Syria; an overview of its long history is available in Reference[5]. Archaeologists have highlighted the emergence of urban societies in the surrounding mountains during the course of the fifth millennium BC. Over the forth millennium BC, the first sophisticated urban society grew up in south Mesopotamia. It was in this region, perhaps in the large city of Uruk on the Euphrates, that writing begun around 3100 BC[6]. Mesopotamia is considered, in particular, as the land where not only writing, but also the wheel, bureaucracy and state have been invented.

---

[3] See the recent study by Eleanor Robson on the social history of mathematics from ancient Iraq [28]. For a summary of the main ideas developed in this work, see [23].

[4] See the recent study on Hammurabi of Babylon by Dominique Charpin [6].

[5] For an overview of the ancient Near East history, [31].

[6] For the invention of cuneiform writing, see [13].

What about mathematics? It is usually assumed by mathematicians that Babylonians:

i) used a sexagesimal place value system (base 60) for counting and calculation, which brought to mind the modern divisions of the circle and the hour;

ii) worked with Pythagorean triples according to the famous tablet Plimpton 322 and extracted square roots;

iii) were able to solve linear systems and cubic equations, but their geometry was sometimes incorrect and not really developed;

iv) knew Pythagoras theorem a millennium before the time at which it is assumed to have been invented by Pythagoras[7].

The reason why these practices and conceptualisations are commonly called *Babylonian* mathematics is that most of the Mesopotamian mathematics known by the general public actually comes from the so-called Old Babylonian period (roughly 1800-1600 BC). We possess several hundred Old Babylonian mathematical tablets, and it is precisely this form of mathematics that historians in general have considered as being "proto-Greek". But, almost any of Babylonian mathematics come actually from Babylon itself; excavating the city from the second millennium has become impossible because it is situated under the water table.

In fact, the various documents providing convincing evidence of mathematics in Mesopotamia come mainly from two widely separated time spans: the Old Babylonian period and the period from around 600 BC to the first century BC[8]. Historians have often ignored such evidence of mathematics dated back to the second period because they considered that mathematics consisted almost exclusively of mathematical procedures and calculations dedicated to astronomy. But, the discovery of few tablets containing mathematical problems with no relation to astronomy has shown that this assertion is not strictly true. In fact, it seems that, although the terminology and external features have changed since the Old Babylonian times, mathematical traditions of the early second millennium have not died out.

Thus, the cuneiform mathematical corpus is more varied and much richer than it has generally been supposed. It ensues that, even though using the expression *Babylonian* is convenient to describe this corpus, this term is unable to exactly and totally reflect the corpus features over more than two thousands years.

---

[7] See, for example [32].

[8] After 1600 BC, mathematical activity appeared to come to a halt in Mesopotamia. But our picture of Mesopotamian mathematics is skewed by the accidents of discovery and depends on excavations. Nevertheless, one knows that, between 1600 and 1000 BCE, mathematical and metrological texts continued to be copied and learnt by apprentice scribes (for example in Aššur on the Tigris or in Hazor on the western coast).

1.2 Sources and methods

The first mathematical cuneiform texts were published in the years before the First World War. Though it was a few decades after the first decipherments of the two main languages using the cuneiform script[9], Akkadian and Sumerian, no true interpretation was made available. The cuneiform script is composed of wedge-shaped impressions on clay tablets and runs horizontally from left to right.

Over the following decades, François Thureau-Dangin and Otto Neugebauer opened the field of *Babylonian mathematics* with the publication of *Textes mathématiques de Suse* and *Mathematische Keilschrifttexte* in the 1930'. As the historical and cultural contexts of both texts were poorly known at that time[10], the internal mathematical features in the tablets were explored by scholars through reconstruction of the techniques and procedures in use[11].

After the Second World War, the methodology in use for working and editing texts has been followed by historians of cuneiform mathematics[12]. In particular, the mathematical content of the texts were translated into a modern language as well as rewritten in the modern mathematical idiom of symbolic algebra. The aim was to, first, approach the ancient calculation techniques and, second, to compare them with the content and concepts of Greek mathematics. Cuneiformists, on the other hand, paid little attention to Babylonian mathematics that was considered to be too technical and only interpretable by mathematicians.

In the 1970s and 1980s, the attention of scholars, such as Marvin Powell and Jöran Friberg, turned their attention to arithmetic and metrology of the third millennium BC, and Denise Schmandt-Besserat developed her theory on the origins of writing and accounting, which were closely intertwined[13]. The interpretation of this early numeracy and metrology needed now to take into account the cultural, political and socio-economic context. This trend led to a revolution of our understanding of ancient mathematical techniques, with the investigations on the language of Old Babylonian algebra carried out by Jens Høyrup and published in 1990[14]. This study was mainly focused on the Akkadian word problems used in mathematical texts and showed that the underlined algebra is based on a *naïve geometry* with a concrete conception of numbers as measured line and area; this method gave an equal weight to numbers and words in text. In order to illustrate it, let us compare two different translations of the same Old Babylonian problem on the difference between the length and width of a rectangle.

---

[9] See the historiography available in [15] and [24].

[10] One should note that, given the lack of archaeological context for most tablets, placing them in an accurate historical context was difficult.

[11] See the remarks in [15] and [28], 268-274.

[12] See [18] and [1].

[13] See, for example, [20], [21], [9], [11] and [29]

[14] See [14].

Two translations of the Old Babylonian Problem AO 8892[15]:

van der Waerden's Translation ([32], 63)	Høyrup's Translation ([16], 164-5)
Length, width. I have multiplied length and width, thus obtaining the area. Then I added to the area, the excess of the length over the width: 3 03 (i.e., 183 was the result). Moreover, I have added the length and width: 27. Required length, width, and area. (given:) 27 and 3 03, the sums (result:) 15 length 3 00, area 12 width One follows this method: 27 + 3 03 = 3 30 2 + 27 = 29 Take one half of 29 (this gives 14;30). 14;30 + 14;30 = 3 30;15 3 30;15 + 3 30 = 0;15 The square root of 0;15 is 0;30. 14;30 + 0;30 = 15 length 14;30 + 0;30 = 14 width. Subtract 2, which has been added to 27, from 14, the width. 12 is the actual width. I have multiplied 15 length by 12 width. 15 + 12 = 3 00 area 15 + 12 = 3 3 00 + 3 = 3 03.	Length, width. I have made length and width hold each other. I have built a surface. I turned around (it). As much as length went beyond width, I have appended to inside the surface: 3 03. I turned back. I have accumulated length and width: 27. What are the length, width, and surface? 27 3 03 the things accumulated 15 the length 12 the width 3 00 the surface You, by your proceeding, append 27, the things accumulated, length and width, to inside 3 03: 3 30. Append 2 to 2: 29. You break its moiety, that of 29: 14;30 steps of 14;30 is 3 30;15. From inside 3 30;15 you tear out 3 30: 0;15, the remainder. The equal-side of 0;15 is 0;30. You append 0;30 to one 14;30: 15, the length. You tear out 0;30 from the second 14;30: 12, the true width. I have made 15, the length, and 12, the width, hold each other: 15 steps of 12 is 3 00, the surface. By what does 15, the length, go beyond 12, the width? It goes beyond by 3. Append 3 to inside 3 00, the surface: 3 03, the surface.

The former is the translation proposed by van der Waerden in 1954, where the problem statement is formulated through two algebraic equations with two unknowns, which can be written as follows:

$$xy + x - y = 183$$

and

$$x + y = 2.$$

The second interpretation, which was proposed by Jens Høyrup, gives special dimension and signification to numbers. They are considered as geometric lengths or areas, which can be manipulated physically (*accumulated*, *appended*, etc.). Høyrup,

[15] I give here an example, which was already described in details by Robson, because it is clear and convenient (for further commentary, see [28], 276-277). All the numbers are given in the modern transliteration of the sexagesimal place value system. Sexagesimal places are separated by a space. In order to approach the absolute value of a sexagesimal number, academics use a semicolon to mark the boundary between the whole and fractional parts of the number. For example, 14; 30 stands for $14 + 30 \times 60^{-1} = 141/2$ and 3 03 for $3 \times 60 + 3 = 183$.

thus, underlined a kind of implicit *cut-and-paste diagram*, contrary to the interpretation by van der Waerden of an ancient algebra with equations and symbols like the one in use nowadays.

An interdisciplinary picture based on this new trend has developed over the last few decades and from the perspective of the social and cultural history of Mesopotamia. Nowadays, historians of cuneiform mathematics try to recover the mathematical thinking and concept of the ancient Mesopotamians from thorough studies of the internal features of mathematical texts (original vocabulary, syntax, etc.) and art factual details of the tablets themselves (visual properties, structure, layout, etc.).

Although it is still very difficult to identify the work of influence of ancient individuals within this almost completely anonymous mathematical tradition, it is only recently that gaining more insight into the people who have used, learned and transmitted mathematical skills and techniques has become possible; let us cite, for example, the recent contribution by Eleanor Robson where she sketched a big picture of the social and cultural history of cuneiform mathematics and opened a unique window onto the material, social, and intellectual world of the mathematics in ancient Iraq[16].

These new perspectives enable one to rethink the relationship between Babylonian and Greek mathematics by interpreting the former as something other than the precursor to the latter. On the one hand, *Babylonian mathematics* seems to be entirely inductive and based on a repetition of specific problems used as generic examples and from which generalisations and conceptualisation are inferred. On the other hand, Greek mathematics is deductive and gives explicitly stated theorems and axioms. Old Babylonian mathematics is based on metric features unlike the classical Greek tradition, which is inherently geometric.

The entire published corpus of cuneiform mathematics comprises now more than 950 tablets. Over 80% of them deal with mathematical and metrological exercises and tables from the Old Babylonian period.

2 Different forms of mathematical thinking and practices

Rather than giving in this paper a likely too short and maybe imprecise overview of cuneiform mathematics it seemed to me worthwhile to focus on two case-studies, both connected with arithmetical and mathematical practices, but over one thousand years distant from each other, so as to describe different way of using and thinking mathematics.

---

[16] [28], 290.

2.1 Counting and accounting around 3000 BC

Since the beginning of writing in southern Mesopotamia, numeracy and literacy have been closely intertwined[17]. It even seems that numeracy has predated literacy by several centuries in Mesopotamia. From about 6000 BC, that is to say several thousand years before writing, Neolithic villagers used geometrically shaped clay or stone counters (or *tokens*) to represent fixed quantities of goods so as to record exchange transactions. Later, in the early fourth millennium, administrators and bureaucrats of the first city states developed systematic accounting techniques by adapting these counters to their increasingly complex needs.

These new practices led to an important change: the use of permanent impressions of the counters on clay in place of ephemeral groups of counters. Whereas the clay counters, or their impression, represented both a number and an object – in other words, a counter had both a quantitative and a qualitative value - the invention of writing around 3200 BC enabled a visual, and certainly conceptual[18], separation of the counting system from the objects to be counted: as a result, the recording numbers were, thus, impressed on clay with a round reed, and the drawing of objects was incised with a sharp reed. These two kinds of writing have evolved from increasing needs of local administrations and bureaucracies for management of goods.

A consequence of this evolution is the possibility to record metric data as well as arithmetic operations, like the sums of counted objects or the total of measured products. Whereas the identification of most of the commodities being counted is still uneasy because of the difficult interpretation of the earliest signs, the comparison of totals and sub-totals of quantities written on many tablets has led to a better understanding of the counting and measuring systems. In 1990's, an interdisciplinary team in Berlin identified more than twelve different systems used on the *archaic* tablets from Uruk, a large city of around 250 hectares in size, located in southern Mesopotamia[19]. Each system depended on the products to be counted or the quantities to be represented. For example, a given system, *i.e.* a set of units, was used to count discrete things such as containers, fish or human beings; another one was employed to count discrete grain products and cheese. About the systems used for noting down capacity measures, each of them depended on specific cereal products (such as barley, kinds of emmer, malt). Other specific systems were dedicated, respectively, to area measurements or to the recording of weights, etc.

One should note that these counting and measuring systems followed strict arithmetic rules. Rigid relationships existed between the units in each system. Depending on the context, a certain number of signs was always replaced by a higher unit. In this respect, most of the numerical and metrological signs changed their arithmetical value according to their field of application. For example, the arithmetic

---

[17] See [27].

[18] See [2].

[19] See [19].

relationships of the signs ● and ▷ impressed in clay, depended on the context.

$$1 \quad ● = 10 \quad ▷ \text{ in context with sheep}$$
$$1 \quad ● = 6 \quad ▷ \text{ in context with barley}$$
$$1 \quad ● = 18 \quad ▷ \text{ in context with field}$$

During the late fourth to the early third millennium, the administrators developed, thus, simultaneously cuneiform script and arithmetic in order to register palace and temple stocks. The invention of the former had consequence for the latter: it enabled one to express a numerical total of different entries on the obverse of a tablet. For example, on the observe of the tablet denoted by W 20274,37 (Vorderasiatisches Museum, Berlin) the sign ▷ appears in seven entries, and each sign represents a jar filled with beer. On the reverse of the tablet, the seven signs are written again, with the sign for "jar" representing the total of the registered jars.

This finding allows one to define, for the first time, this kind of operation as *calculation* or *addition*, even though this addition was not an arithmetical operation in the proper sense of the word; it, indeed, corresponds to a kind of manipulation of the counted objects. But this principle was actually complex, because repeated signs within each system were replaced with larger units: for example, the sum of twelve signs ▷ for counting animals would have been written ● ▷ ▷ and followed with the sign corresponding to *sheep*.

Even though this activity should not be called mathematics, this "proto-arithmetic", with the expressed and real summarising process, formed the point of departure of an abstraction concerning symbolic operation with numbers, which does not depend on counted objects[20].

2.2 Elementary Mathematics education in the Early Second Millennium

Around one thousand years later, there was in Mesopotamia a highly centralized bureaucratic state during the so-called Ur-III period, with its enormous bureaucratic apparatus. In order to form scribes and provide them with administrative training, a system of standardised schools was created in the state.

Unfortunately, little is known about this kind of scribal education. A hymn of the king Shulgi (2094-2047 BC), a famous king of the Ur-III period gives us some clues:

When I was young, I learnt at school the scribal art on the tablets of Sumer and Akkad. Among the high born no-one could write like me. Where people go for instruction in the scribal art there, I mastered completely subtraction, addition, calculation and accounting. (Hymne B l.13-20: see [34], 24-25).

---

[20] See [7] for this development of the number concept.

As already seen, most of the excavated mathematical tablets dated back to the so-called Old Babylonian period, at the early second millennium. After the implosion of the Ur-III Empire, which had collapsed under its too heavy bureaucracy, Mesopotamia was divided into small city-states, with their own administrative and economic functions. Private entrepreneurs played an important role in these local administrations.

These economic centres still needed scribes for accounting and stock management, and many thousands of school tablets have been excavated. Although most of them are from an unknown provenance, archaeologists have, nevertheless, found mathematical tablets in some sites such as Mari by the Euphrates or Susa in southwest Iran, as well as several school houses from southern Irak[21].

Mathematics was part of the scribal curriculum, which also included grammar, literature and good practice in the writing of legal documents, letters and accounts; the Old Babylonian education system was targeted to the production of numerate and literate scribes[22].

The Old Babylonian mathematical texts are usually categorised as *problems, tables* (usually including metrological lists), *round tablets for calculations, solution lists* and *coefficient lists* in the modern literature. The very first text was usually composed and transmitted by the scribal teachers, and further copied by the apprentice scribes. They contained series of problems with their numerical answers, which consisted in some instructions expressed in the imperative form, or in the second-person singular, often written in Akkadian (see the Old Babylonian Problem AO 8892 above) and ending with the expression *this is the procedure*. Series of problems often differed minimally: indeed, the procedure was alike, and they only contained alternative sets of parameters, in a pedagogical order.

Students wrote out tables, often with Sumerian words; they consisted in multiplication tables and reciprocal tables (pair of numbers written in sexagesimal place value system with their inverses). For example, Ashmolean 1924.447 is a multiplication table for 24:[23]

Obverse

1. steps of 16	6 24	
steps of 17	6 48	
steps of 18	7 12	
steps of 20-lá-1	7 36	
5. steps of 20	8	
steps of 6	2 24	
steps of 7	2 48	
steps of 8	3 12	
steps of 9	3 36	

---

[21] See [5], 419-33.

[22] For this curriculum, see [25], [28], 85-113 and [33]. An example of mathematical education at Ur is given [12].

[23] See [26].

10. steps of 10   4
    steps of 11   4 24
    steps of 12   4 48
    steps of 13   5 12
    steps of 14   5 36
15. steps of 15   6
    Reverse
1. steps of 16   6 24
    steps of 17   ☐6 48☐
    steps of 18   7 12
    steps of 20-lá-1   7 36
5. steps of 20   ☐8☐
    steps of 30   12
    steps of 40   16
    steps of 50   20
    (*erasures*)
10. 'Long tablet' of (the scribe) Suen-apil-Urim
    Month XII, day 9
    Upper edge
    Praise (the gods) Nisaba (and) Ea!

Apprentice scribes had to copy and memorise the so-called metrological lists, which enumerate measures of capacity, weight, surface and length in increasing order. The pedagogical aims of this work were twofold; the lists provided information about the units in use in each metrological system and their notations[24]. Their vertical enumeration enabled trainee scribes to express and to memorise the ratios between the units in each system. It can be seen, for instance, in a sequence from the weight system. The numerical signs u (diagonal wedge) and diš (vertical wedge) were used to express 10 and 1, respectively. The signs for the weight measures were, in order of size, gin_2 for "shekel" (c . 8 g) and ma-na for "mina" (c. 500 g):

Transliteration	Translation
1(u) 1(diš) gin_2	11 shekels
1(u) 2(diš) gin_2	12 shekels
1(u) 3(diš) gin_2	13 shekels
1(u) 4(diš) gin_2	14 shekels
1(u) 5(diš) gin_2	15 shekels
1(u) 6(diš) gin_2	16 shekels
1(u) 7(diš) gin_2	17 shekels
1(u) 8(diš) gin_2	18 shekels
1(u) 9(diš) gin_2	19 shekels

---

[24] See comments in [28], 96-106, [3] and [4].

1/3 ma-na	1/3 mina
1/2 ma-na	1/2 mina
2/3 ma-na	2/3 mina
5/6 ma-na	5/6 mina
1 ma-na	1 mina

The ratio between the two units is expressed with fractional values: the notation "1/3 mina" directly follows the increasing progression "11 shekels, 12 shekels, 13 shekels . . . 19 shekels", indicating that "1/3 mina" is equal to "20 shekels", and, thus, that "1 mina" is equal to 60 shekels.

With these tables, scribes performed calculations in the sexagesimal place value system on small round tablets, called *hand tablets* in Sumerian. Solution lists, with sets of parameters giving integer answers for individual problems, were kept by the teachers[25].

Finally, technical constants – work-rates and standards already in use in the Ur-III period – were organised into lists in order to solve problems[26].

Although mathematical problems seem to be based on practical needs (measurements of grain-piles, calculations for earthworks and waterworks, and for manufacturing standardised bricks, etc.), the non-utilitarian data of some problems suggest that Old Babylonian mathematics was concerned with approximation of the real world, without tempting to be "abstract mathematics" focused on mathematical modelling.

3 Conclusions

This paper attempted to highlight the interactions between ancient mathematics and society. It is now incorrect to consider *Babylonian mathematics* as the precursor to deductive Greek mathematics. *Babylonian mathematics* had its own features, which were products of the daily life needs for administration, economy and politics, and more generally of the society that invented, used and transmitted them.

Wherever possible the interpretation of a mathematical text thus needs to take into account both the context of its drafting (layout of the document, vocabulary, scribal traditions, etc.) and the context of actual practice (surveying fields, trading and pilling up grain, methods of accounting, etc.), rather than focusing only on the mathematical content of the text. The recent studies on mathematics in Mesopotamia have shown that this contextualisation has now become a guiding thread enabling one to evidence that the mathematical cuneiform corpus is more varied and complex than it had usually been acknowledged.

To sum up, the study of the ancient cultures, which have produced and expressed mathematics, has become necessary and gives rise to new interdisciplinary cooperation between mathematicians, historians of mathematics and archaeologists.

---

[25] See [8].

[26] See [22].

References

1. E.M. Bruins, M. Rutten, Textes mathématiques de Suse (Mémoires de la Mission Archéologique en Iran 34). Geuthner, Paris, 1961.
2. E. Cancik-Kirschbaum, G. Chambon, Maβangaben und Zahlvorstellung in archaischen Texten. Altorientalische Forschungen **3**: 189-214, 2006.
3. G. Chambon, Numeracy and Metrology. In: K. Radner, E. Robson (eds.), Oxford Handbook of Cuneiform Culture. Oxford, 2011.
4. G. Chambon, E. Robson, Untouchable or unrepeatable? The upper end of the Old Babylonian metrological systems for capacity and area. Iraq **73**: 127-147, 2011.
5. D. Charpin, Le clergé d'Ur au siècle d'Hammurabi (XIXe–XVIIIe siècles av.J.-C.) (Hautes Études Orientales 22). Droz, Geneva, 1986.
6. D. Charpin, Hammu-rabi de Babylone. PUF, Paris, 2003.
7. P. Damerow, Vorüberlegungen zu einer historischen Epistemologie der Zahlbegriffsentwicklung. In: G. Dux, U. Wenzel (eds.), Der Prozess der Geistesgeschichte. Studien zur ontogenetischen und historischen Entwicklung des Geistes, Frankfurt am Main, 248-322, 1994.
8. J. Friberg, Methods and traditions of Babylonian mathematics: Plimpton 322, Pythagorean triples and the Babylonian triangle parameter equations. Historia Mathematica **8**: 277-318, 1981.
9. J. Friberg, Three remarkable texts from ancient Ebla. Vicino Oriente **6**: 3-25, 1986.
10. J. Friberg, Mathematik. In: Reallexikon der Assyriologie, vol. 7, ed. D.O. Edzard, 457-517. de Gruyter, Berlin, 1987-90.
11. J. Friberg, Round and almost round numbers in proto-literate metromathematical field texts. Archiv für Orientforschung **44-5**: 1-58, 1997-98.
12. J. Friberg, Mathematics at Ur in the Old Babylonian period. Revue d'Assyriologie **94**: 97-188, 2000.
13. J.-J. Glassner, The invention of cuneiform: writing in Sumer, trans. Z. Bahrani and M. Van De Mieroop. John Hopkins University Press, Baltimore, 2003.
14. J. Høyrup, Algebra and naive geometry. An investigation of some basic aspects of Old Babylonian mathematical thought. Altorientalische Forschungen **17**: 27–69, 262–354, 1990.
15. J. Høyrup, Changing trends in the historiography of Mesopotamian mathematics: an insider's view. History of Science 34: 1-32, 1996.
16. J. Høyrup, Lengths, widths, surfaces: a portrait of Old Babylonian algebra and its kin. Springer, Berlin, 2002.
17. O. Neugebauer, Mathematische Keilschrift - Texte, 3 vols. (Quellen und Studien zur Geschichte der Mathematik, Astronomie und Physik A3). Springer, Berlin, 1935-37.
18. O. Neugebauer, A.J. Sachs, Mathematical Cuneiform Texts (American Oriental Series 29). American Oriental Society, New Haven, 1945.
19. H.J. Nissen, P. Damerow, R.K. Englund, Archaic bookkeeping: early writing and techniques of administration in the ancient Near East. Chicago University Press, Chicago, 1993.
20. M.A. Powell, The antecedents of Old Babylonian place notation and the early history of Babylonian mathematics. Historia Mathematica **3**: 417-39, 1976.
21. M.A. Powell, Masse und Gewichte. In: Reallexikon der Assyriologie, vol. 7, ed. D.O. Edzard, 457-517. de Gruyter, Berlin, 1987-90.
22. E. Robson, Mesopotamian mathematics, 2100–1600 BC: technical constants in bureaucracy and education (Oxford Editions of Cuneiform Texts 14). Clarendon, Oxford, 1999.
23. E. Robson, Mesopotamian mathematics: some historical background. In: V.J. Katz (ed.), Using History to Teach Mathematics. Mathematical Association of America, Washington DC, 2000, 149-158, 2000.
24. E. Robson, Guaranteed genuine Babylonian originals: the Plimpton collection and the early history of mathematical Assyriology. In: C. Wunsch (ed.), Mining the archives: Festschrift for C.B.F. Walker, 245–92. ISLET, Dresden, 2002.

25. E. Robson, More than metrology: mathematics education in an Old Babylonian scribal school. In: J.M. Steele, A. Imhausen (eds.), Under one sky: mathematics and astronomy in the ancient Near East (Alter Orient und Altes Testament 297), 325-65. Ugarit-Verlag, Münster, 2002.
26. E. Robson, Mathematical cuneiform tablets in the Ashmolean Museum, Oxford. SCIAMVS—Sources and Commentaries in Exact Sciences 5: 3–65, 2004.
27. E. Robson, Literacy, numeracy, and the state in early Mesopotamia. In: K. Lomas, R.D. White-house, J.B. Wilkins (eds.), Literacy and the State in the Ancient Mediterranean. Accordia Research Institute, London, 37-50, 2007.
28. E. Robson, Mathematics in Ancient Iraq: a Social History. Princeton University Press, 2008.
29. D. Schmandt-Besserat, Before writing: from counting to cuneiform. University of Texas Press, Austin, 1992.
30. F. Thureau-Dangin, Textes mathématiques babyloniens (Ex Oriente Lux 1). Brill, Leiden, 1938.
31. M. van De Mieroop, A history of the ancient Near East, c. 3000–323 BC. Blackwell, Oxford, 2004.
32. B.L. Van der Waerden, Science awakening, trans. A Dresden. Noordhoff, Groningen, 1954.
33. N. Veldhuis, How did they learn cuneiform? Tribute/Word List C as an elementary exercise. In: Approaches to Sumerian literature: studies in honor of Stip (H.L.J. Vanstiphout), P. Michalowski, N. Veldhuis (eds.), 181-200. Brill, Leiden, 2006.
34. N. Veldhuis, Elementary Education at Nippur. Groningen, 1997.

Origami and Partial Differential Equations

Paolo Marcellini and Emanuele Paolini

Origami is the ancient Japanese art of folding paper and it has well known algebraic
and geometrical properties, but it also has unexpected relations with partial differential equations. In this note we describe these relations for a large audience, leaving
the technical aspects to other specialized papers.

Introduction

Origami is the ancient Japanese art of folding paper. One of the most known origami
is the *crane*, represented on the right-hand side of Fig. 1. Other than their artistic
interest, why and how to associate origami with mathematics?

A motivation comes from the properties of origami. Many mathematicians interested in geometry or algebra (for example in group theory, Galois theory, graph
theory) studied origami constructions.

An important issue is the geometrical construction of numbers. In some aspect
origami turns out to be more powerful than the classical *rule and compass* construction. In fact, in order to determine what can be constructed through origami, it is
important to formalize the rules. These are known as *Huzita axioms* and have been
proposed by Hatori, Huzita, Justin and Lang, see [1].

On the contrary, in this exposition we present an analytic approach to origami,
based on maps which satisfy a suitable system of partial differential equations. We
remain here to a non-technical level of exposition. The interested reader might refer
to the papers [5,7] obtained by the authors in collaboration with Bernard Dacorogna
(*École Polytechnique Fédérale de Lausanne*). In these papers fractal constructions
of origami are shown to solve a special class of Dirichlet problems arising in nonlinear elasticity (see [4]).

Paolo Marcellini and Emanuele Paolini
Department of Mathematics, University of Firenze (Italy).

Emmer M. (Ed.): Imagine Math. Between Culture and Mathematics
DOI 10.1007/978-88-470-2427-4_22, © Springer-Verlag Italia 2012

1 Axiomatic construction of origami

As already said in the introduction, origami constructions can be considered from an axiomatic point of view in a similar way as rule and compass construction. We give here few details about this geometric approach to origami (the interested reader may refer to [1]).

Here are the *seven axioms*.

- *Axiom 1:* given two points P_1 and P_2, there is a unique fold passing through both of them;
- *Axiom 2:* given two points P_1 and P_2, there is a unique fold placing P_1 onto P_2;
- *Axiom 3:* given two lines L_1 and L_2, there is a fold placing L_1 onto L_2;
- *Axiom 4:* given a point P and a line L, there is a unique fold perpendicular to L passing through P;
- *Axiom 5:* given two points P_1 and P_2 and a line L, there is a fold placing P_1 onto L and passing through P_2;
- *Axiom 6:* given two points P_1 and P_2 and two lines L_1 and L_2, there is a fold placing P_1 onto L_1 and P_2 onto L_2;
- *Axiom 7:* given a point P and two lines L_1 and L_2, there is a fold placing P onto L_1 and perpendicular to L_2.

However this is not the only possible mathematical motivation and in the following we propose a different approach. We will present a mathematical model of origami which has a double purpose. In one hand we give an *analytical approach* which provides a new perspective to the existing algebraic and geometrical models. In the other hand we use origami as a tool to exhibit explicit solutions to some systems of partial differential equations.

2 A global definition of origami as a map

Instead of listing a set of properties, we identify an origami with a mathematical object, i.e., we give a mathematical model. We skip the *overlapping* and *interpenetration* problems (see [5]).

If we denote by $\Omega \subset \mathbb{R}^2$ a two dimensional domain (usually Ω is a rectangle), then an origami is a *suitable* immersion of the sheet of paper in the three dimensional space. Hence it can be identified with a map u

$$u\colon \Omega \subset \mathbb{R}^2 \to \mathbb{R}^3.$$

Since origami is a folded paper, the map u cannot be everywhere smooth; it is only piecewise smooth. In fact *folding creates discontinuities in the gradient*. But we do not allow cutting the sheet of paper. Thus u is a continuous map.

The *singular set* $\Sigma = \Sigma_u$ is the set of discontinuities of the gradient Du. This set represents the union of curves where the paper is folded and hence it is also

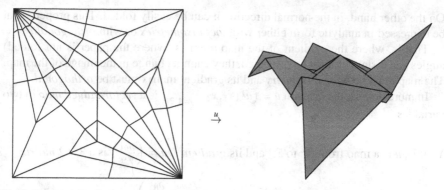

Fig. 1 On the right: the crane is the most famous origami. On the left: the corresponding singular set

called *crease pattern* in the origami context. Usually this set is composed by straight segments.

In the model – at the same time – we construct the origami and we unfold it. We now explain in which sense.

Let's consider the crane origami represented in Fig. 1. If we unfold the origami we see the crease pattern Σ impressed in the sheet of paper. Clearly *the singular set Σ is uniquely determined* by the origami. In this case Σ is *the set of segments* represented on the left in Fig. 1. What we really consider is a function, an application, a map u from *the sheet of paper* to *the three-dimensional space*.

As we said, usually the singular set is composed by straight segments, but it is also possible to make origami with curved folds: this happens for instance in the representation of a map $u: \mathbb{R}^2 \to \mathbb{R}^3$, which has as singular set Σ_u along a circular curve, as in Fig. 2.

A sheet of paper Ω (again recall that usually Ω is a rectangle) is rigid in tangential directions. If a sheet of paper is constrained on a plane, it would only be possible to achieve rigid motions, i.e., rotations and translations of the whole sheet.

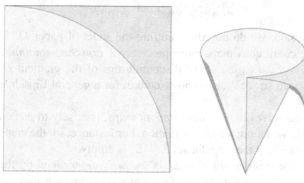

Fig. 2 A non-flat origami with a curved singular set

On the other hand, in the normal direction it can be easily folded. This property can be expressed in analytic form either with *local isometries* or with *orthogonality*.

That is, where the gradient of the map u exists (where the paper is not folded) angles and distances must be respected, they cannot change in the image of the map. The map must be a *local isometry* and its gradient matrix must be *orthogonal*.

In more details, the origami $u = \left(u^j\left(x_1, x_2\right)\right)_{j=1,2,3}$ is a *vector-valued map* in two variables

$$u \colon \mathbb{R}^2 \to \mathbb{R}^3.$$

That is, u is a map from $\mathbb{R}^2$ to $\mathbb{R}^3$ and its *gradient* $Du = \left(\frac{\partial u^j}{\partial x_i}\right)$ is a 3×2 *matrix*

$$u = \begin{pmatrix} u^1 \\ u^2 \\ u^3 \end{pmatrix}, \quad Du = \begin{pmatrix} \frac{\partial u^1}{\partial x_1} & \frac{\partial u^1}{\partial x_2} \\ \frac{\partial u^2}{\partial x_1} & \frac{\partial u^2}{\partial x_2} \\ \frac{\partial u^3}{\partial x_1} & \frac{\partial u^3}{\partial x_2} \end{pmatrix}.$$

The gradient $Du(x)$ has to be an *orthogonal* 3×2 matrix, i.e.,

$$Du^t \cdot Du = I.$$

This orthogonality condition is equivalent to the *differential system*

$$\sum_{i=1}^{3} \frac{\partial u^i}{\partial x_h} \cdot \frac{\partial u^i}{\partial x_k} = \delta_{hk}, \quad \forall\, h, k = 1, 2,$$

which, in explicit form, means

$$\begin{cases} \left(\frac{\partial u^1}{\partial x_1}\right)^2 + \left(\frac{\partial u^2}{\partial x_1}\right)^2 + \left(\frac{\partial u^3}{\partial x_1}\right)^2 = 1 \\[2mm] \left(\frac{\partial u^1}{\partial x_2}\right)^2 + \left(\frac{\partial u^2}{\partial x_2}\right)^2 + \left(\frac{\partial u^3}{\partial x_2}\right)^2 = 1 \\[2mm] \frac{\partial u^1}{\partial x_1}\frac{\partial u^1}{\partial x_2} + \frac{\partial u^2}{\partial x_1}\frac{\partial u^2}{\partial x_2} + \frac{\partial u^3}{\partial x_1}\frac{\partial u^3}{\partial x_2} = 0. \end{cases}$$

As we already said, we do not allow cutting the sheet of paper Ω. Thus $u \colon \Omega \subset \mathbb{R}^2 \to \mathbb{R}^3$ is a continuous map, more precisely a *Lipschitz-continuous map*. The singular set $\Sigma = \Sigma_u$, i.e., the set of discontinuities of the gradient Du, may have a very complicated structure, even no-structure, for a general Lipschitz-continuous map.

If we limit ourselves to *piecewise smooth maps*, precisely to *piecewise C^1 rigid maps*, then we have a more readable situation. For instance, for the map whose graph is represented in Fig. 3 the singular set $\Sigma = \Sigma_u$ is empty.

As we said the singular set $\Sigma = \Sigma_u$ is *uniquely determined* by the map u, but in general the reverse is not true; in fact many rigid maps u may have the same singular set. On the contrary a special attention will be given to the so-called flat

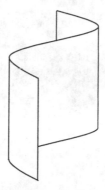

Fig. 3 This sheet of paper $u(\Omega)$ is bended but not folded. The corresponding singular set Σ_u is empty (in correspondence to several (not folded) maps)

origami. A *flat origami* is defined as a map whose image is contained in a plane. It can be represented, up to a change of coordinates, as a map $u\colon \Omega \subset \mathbb{R}^2 \to \mathbb{R}^2$. Let us consider a *flat origami*, i.e., instead of

$$u\colon \Omega \subset \mathbb{R}^2 \to u(\Omega) \subset \mathbb{R}^3,$$

we consider an application of the form

$$u\colon \Omega \subset \mathbb{R}^2 \to u(\Omega) \subset \mathbb{R}^2.$$

3 Analytic properties of flat origami

In the case of flat origami we have the possibility of reconstructing the map u from its singular set Σ_u. That is, if $u(\Omega) \subset \mathbb{R}^2$, it is possible to *uniquely reconstruct* a map, with orthogonal gradient, from a given set of singularities; i.e., from a given singular set. A fundamental ingredient in this reconstruction is a necessary and sufficient *compatibility condition* on the geometry of the singular set.

Following the terminology that can be found in the not numerous mathematical literature on origami (see for instance [2]), we call it *angle condition*. It was discovered by Kawasaki in the origami setting.

Let $\Sigma \subset \Omega \subset \mathbb{R}^2$ be a locally finite union of segments. Then Σ is the *singular set* of a *piecewise C^1 rigid map* if and only if the following *angle condition* holds at every *internal* vertex of Σ. If we let $\alpha_1, \ldots, \alpha_N$ be the *amplitude of the consecutive angles* determined by the N edges of Σ meeting in the vertex, then N is even and (see Fig. 4)

$$\alpha_1 + \alpha_3 + \ldots + \alpha_{N-1} = \alpha_2 + \alpha_4 + \ldots + \alpha_N = \pi.$$

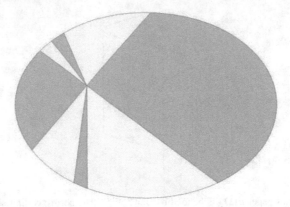

Fig. 4 *the angle condition*: at every internal vertex an even number of angles meet. The alternating sum of angles is equal each other

We prove that every polyhedral pattern Σ which satisfies the *angle condition* is the singular set Σ_u of some rigid map u. Precisely the following result holds (the result is valid in the general n–dimensional setting, with $\Omega \subset \mathbb{R}^n$).

The following result has been proved in [5].

Theorem 1 (Recovery Theorem). *Let Ω be a simply connected open subset of $\mathbb{R}^2$. Let $\Sigma \subset \Omega$ be a locally finite polyhedral set satisfying the angle condition at every vertex. Then there exists a map u with orthogonal gradient (flat origami) such that $\Sigma = \Sigma_u$ is the singular set of u. Moreover u is uniquely determined once we fix the value $y_0 = u(x_0)$ and the Jacobian gradient $J_0 = Du(x_0)$ at a point $x_0 \in \Omega \setminus \Sigma$.*

For a *flat origami* $u \colon \Omega \subset \mathbb{R}^2 \to u(\Omega) \subset \mathbb{R}^2$, with components (with a little abuse of notation we identify $\mathbb{R}^2$ with a subset of $\mathbb{R}^3$)

$$u = \begin{pmatrix} u^1 \\ u^2 \\ 0 \end{pmatrix} = \begin{pmatrix} u^1 \\ u^2 \end{pmatrix}, \quad Du = \begin{pmatrix} \frac{\partial u^1}{\partial x_1} & \frac{\partial u^1}{\partial x_2} \\ \frac{\partial u^2}{\partial x_1} & \frac{\partial u^2}{\partial x_2} \\ 0 & 0 \end{pmatrix} = \begin{pmatrix} \frac{\partial u^1}{\partial x_1} & \frac{\partial u^1}{\partial x_2} \\ \frac{\partial u^2}{\partial x_1} & \frac{\partial u^2}{\partial x_2} \end{pmatrix},$$

the *orthogonality condition* $Du^t \cdot Du = I$ is equivalent to the *differential system*

$$\begin{cases} \left(\dfrac{\partial u^1}{\partial x_1} \right)^2 + \left(\dfrac{\partial u^2}{\partial x_1} \right)^2 = 1 \\[2mm] \left(\dfrac{\partial u^1}{\partial x_2} \right)^2 + \left(\dfrac{\partial u^2}{\partial x_2} \right)^2 = 1 \\[2mm] \dfrac{\partial u^1}{\partial x_1} \dfrac{\partial u^1}{\partial x_2} + \dfrac{\partial u^2}{\partial x_1} \dfrac{\partial u^2}{\partial x_2} = 0 \end{cases} \tag{1}$$

and this gives a representation for the *determinant* of the 2×2 matrix Du. In fact, by an algebraic computation, we also find

$$(\det Du)^2 = \left(\frac{\partial u^1}{\partial x_1} \frac{\partial u^2}{\partial x_2} - \frac{\partial u^2}{\partial x_1} \frac{\partial u^1}{\partial x_2} \right)^2$$

$$= \left(\frac{\partial u^1}{\partial x_1} \frac{\partial u^2}{\partial x_2} \right)^2 + \left(\frac{\partial u^2}{\partial x_1} \frac{\partial u^1}{\partial x_2} \right)^2 - 2 \frac{\partial u^1}{\partial x_1} \frac{\partial u^2}{\partial x_2} \frac{\partial u^2}{\partial x_1} \frac{\partial u^1}{\partial x_2}.$$

By multiplying side by side the first two equations of (1) we get

$$\left(\frac{\partial u^1}{\partial x_1} \frac{\partial u^1}{\partial x_2} \right)^2 + \left(\frac{\partial u^1}{\partial x_1} \frac{\partial u^2}{\partial x_2} \right)^2 + \left(\frac{\partial u^2}{\partial x_1} \frac{\partial u^1}{\partial x_2} \right)^2 + \left(\frac{\partial u^2}{\partial x_1} \frac{\partial u^2}{\partial x_2} \right)^2 = 1$$

and therefore

$$(\det Du)^2 = 1 - \left(\frac{\partial u^1}{\partial x_1} \frac{\partial u^1}{\partial x_2} \right)^2 - \left(\frac{\partial u^2}{\partial x_1} \frac{\partial u^2}{\partial x_2} \right)^2 - 2 \frac{\partial u^1}{\partial x_1} \frac{\partial u^2}{\partial x_2} \frac{\partial u^2}{\partial x_1} \frac{\partial u^1}{\partial x_2}$$

$$= 1 - \left(\frac{\partial u^1}{\partial x_1} \frac{\partial u^1}{\partial x_2} + \frac{\partial u^2}{\partial x_1} \frac{\partial u^2}{\partial x_2} \right)^2.$$

Then, by the third equation in (1), we finally get

$$\det Du = \pm 1.$$

The sign of the determinant of the matrix Du gives a *coloration* of the domain Ω, as in Fig. 5.

Fig. 5 The domain Ω colored by means of the sign of $\det Du = \pm 1$

4 Boundary value problems and fractal constructions

We have another condition to satisfy: it is a *boundary condition*. That is, we look for maps u with a given value at the boundary $\partial\Omega$ of Ω:

$$u(x) = \varphi(x), \quad x \in \partial\Omega.$$

For instance $u(x) = 0$ for $x \in \partial\Omega$. In order to achieve the boundary datum we must arrive at the boundary with finer and finer subdivisions of the set Ω; i.e., we must have a singular set Σ_u of *fractal form*. This is due to the fact that $\det Du = \pm 1$, in particular $\det Du \neq 0$ and hence, by the implicit function theorem, the map u is locally invertible if it is smooth. This is in contrast with a constant boundary value.

We apply the recovery theorem to a singular set Σ_u of fractal form (at the boundary, with the aim to satisfy a boundary condition), for which the *angle condition* is satisfied. In fact the set Σ_u which we are going to consider has the property that it divides Ω into *two families of colored sets* (see Fig. 6):

- grey *rectangles*, where $\det Du = -1$;
- white *convex polygons*, where $\det Du = +1$.

Each vertex of the singular set Σ_u is shared by two rectangles, hence the *angle condition* holds. We see the shape of the sets that we consider: it is an *Escher-type not-periodic picture*.

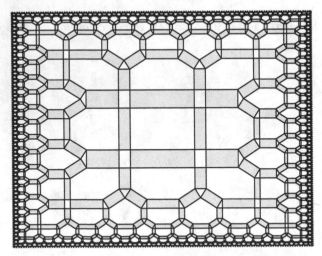

Fig. 6 Escher-type not-periodic picture satisfying the *angle condition* at every vertex, with *fractal structure* at the boundary which allows to fix a *boundary value*

There exists a piecewise C^1 rigid map $u : \bar{\Omega} \to \mathbb{R}^2$ (flat origami), with singular set Σ_u as in Fig. 6, such that $u = \varphi$ on $\partial\Omega$. Thus u satisfies the *Dirichlet problem*

$$\begin{cases} Du \in O(2), & \text{a.e. } x \in \Omega \\ u(x) = \varphi(x), & x \in \partial\Omega \end{cases}$$

for some given boundary values φ (see [5]).

From the *scalar* picture (see Fig. 6) we can also *read* the boundary value of the *vectorial* map u.

We end by giving a picture with a $3-$*dimensional flat origami*. It is a *mathematical origami*, being a rigid application from $\mathbb{R}^3 \to \mathbb{R}^3$.

Theorem 2 (3D Dirichlet Problem). *On the cube* $\Omega = [0,1]^3$ *it is possible to define a piecewise* C^1 *rigid map* $u : \Omega \to \mathbb{R}^3$ *such that* $u = 0$ *on the boundary. The singular set* Σ_u *is represented in Fig. 7.*

This result was first obtained by Cellina and Perrotta [3] and extended in [6] to general n-dimensional origami.

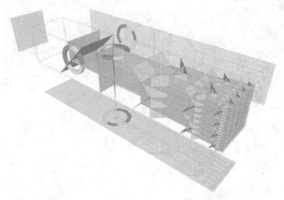

Fig. 7 The singular set which defines a 3-dimensional origami. The angle condition is satisfied on every edge (the rings highlight the measures of the alternating angles)

References

1. R.C. Alperin, A mathematical theory of origami constructions and numbers. J. Math. **6**, 119–133, 2000.
2. M. Bern, B. Hayes, The complexity of flat origami. Proceedings of the 7th Annual ACM-SIAM Symposium on Discrete Algorithms, 175–183, 1996.
3. A. Cellina, S. Perrotta, On a problem of potential wells. J. Convex Analysis **2**, 103–115, 1995.

4. B. Dacorogna, P. Marcellini, Implicit partial differential equations. Progress in Nonlinear Differential Equations and Their Applications, vol. 37. Birkhäuser, Boston, 1999.
5. B. Dacorogna, P. Marcellini, E. Paolini, Lipschitz-continuous local isometric immersions: rigid maps and origami. J. Math. Pures Appl. **90**, 66–81, 2008.
6. B. Dacorogna, P. Marcellini, E. Paolini, On the n-dimensional Dirichlet problem for isometric maps. Journal Functional Analysis **255**, 3274-3280, 2008.
7. B. Dacorogna, P. Marcellini, E. Paolini, Origami and partial differential equations. Notices of AMS **57**, 598–606, 2010.
8. T. Hull, On the mathematics of flat origamis. Congressus Numerantium **100**, 215–224, 1994.
9. T. Kawasaki, On the relation between mountain-creases and valley creases of a flat origami. Proceedings of the 1st International Meeting of Origami Science and Technology, Ferrara, H. Huzita, ed., 1989, 229–237.

Mathematics and Film

The Möbius Strip

a film by Edouard Blondeau

Who is this guy sitting on the couch of the heroine? Why does he punch her in the face? But above all why does the story keep on starting over and over? The Möbius strip is a short film about two main movie characters running away from the situation they have to act in a movie to live their own live.

1 Synopsis

As she wakes up the heroine finds the hero she doesn't know sitting on her couch. She wants him to leave so he stuns her with a punch. The story starts from the beginning again without the heroine and the hero noticing it. So, as she wakes up the heroine finds the hero she doesn't know sitting on her couch. She wants him to leave so he stuns her with a punch.

The heroine wakes up again in her bed. However she has a feeling of "déjà vu". She says it to the hero who doesn't care about it. He stuns her again. For the fourth time the heroine wakes up in her bed. She's irritated by the repetitions and asks the hero for explications. He answers her they are the main characters of a movie but he doesn't know why the story keeps on starting over and over.

So they meet the producer of the movie who explains them that the director doesn't know how to continue the story. That's why he keeps on rehashing the same part over and over and despite himself condemns the heroine and the hero to live indefinitely the same scene. To stop this situation the hero threatens the producer so that he doesn't finance the movie anymore. The producer gives up and tears to pieces the screenplay.

Immediately the heroes are projected on a blank page where they can write their own story thanks to doors opening on different places. So the hero invites the heroine to drink a cocktail on a beach. While they bask in the sun, a director proposes them acting in a movie. They accept.

Whereas the frame remains black until the end, the director calls "action" and we can hear the same conversation between the heroes as in the beginning of the movie.

Edouard Blondeau
Movie director, Paris (France).

Emmer M. (Ed.): Imagine Math. Between Culture and Mathematics
DOI 10.1007/978-88-470-2427-4_23, © Springer-Verlag Italia 2012

2 Director's statement

Through the story of "The Möbius strip" I want to show the imprisonment of the individual in his own background. The individual gives rise to his own barrier (the religion, the social order, other people's look...) what prevents him from evolving. However when he succeeds in crossing these barriers, he often inflicts himself new barriers, and so can be in the quite same situation as before.

To describe the characters obtaining the freedom little by little, the frame becomes gradually bigger until the heroes are meeting the director on the beach. After this moment, the frame becomes smaller: the close shot on the heroes sitting on deckchairs, and then the close-up on the director smiling. The film ends without picture and only the sound suggesting that the story keeps going over and over.

Until the scene with the producer, in order to emphasize the sensation that the character of movie director is short of fantasy the sets are sober with white as main colour and the sound is bared. So the sensation that there is no world outside the story the heroes live, is increased. This fantasy blank is showed also by the neutral personalities of the main characters compared with the charismatic personality of the producer. The only coloured scene takes place on the beach where the sea, the wind and the birds can be heard. As well this is the moment when the heroes are living the life they want. That's what the music enhances with a slight and melodious theme.

The title makes reference to the famous Möbius strip, symbol of infinity that has the particularity of being a 3D geometric form while it has only one surface.

So it is possible to follow endless the strip with the finger on the two sides without crossing the edge. We think we are on a side and then on the other side, but we don't. There is only one side. It's the same progress for the characters who think they are free and finally live again the situation they run away.

3 Script

The short film is also viewable under this web link:
http://www.youtube.com/watch?v=wfkxEsKrOyo

Sequence 1
Sleeping in bed the heroine opens her eyes, seats on the bed and goes out the room.

Sequence 2
The heroine walks to the kitchen. In the living room, she sees a stranger, the hero, sitting on the couch.

The heroine *surprised*: - Hello.
The hero: - Hello.
The heroine: - Who are you? What are doing in my apartment?

The hero: - Ah, so this your place?
The heroine: - Yes it is! So get out of here!

The hero looks around.

The hero: - Nice place.

The heroine walks up to him kick him out.

The heroine: - I don't ask you for opinion! Get the hell out of here!
The hero: - Hey oh!

The hero punches the heroine who falls K.O. on the ground.

Sequence 3
Sleeping in bed the heroine opens her eyes, seats on the bed and goes out the room.

Sequence 4
The heroine walks to the kitchen. In the living room, she sees a stranger, the hero, sitting on the couch.

The heroine *surprised*: - Hello.
The hero: - Hello.
The heroine: - Who are you? What are doing in my apartment?
The hero: - Ah, so this your place?
The heroine: - Yes it is! So get out of here!

The hero looks around.

The hero: - Nice place.

The heroine walks up to him kick him out.

The heroine: - I don't ask you for opinion! Get the hell out of here!
The hero: - Hey oh!

The hero punches the heroine who falls K.O. on the ground.

Sequence 5
Sleeping in bed the heroine opens her eyes remembering something happened.

Sequence 6
The heroine comes in the living room but stays away from the hero.

The heroine: - Tell me... Don't you this feeling of "déjà vu"?
The hero: - No, not really. I'm pretty good with faces and if we'd already met, I'd remember.
The heroine: - No... Don't you have the feeling you and I have already experienced this?
The hero: - ...

The heroine *coming to the hero*: - What I mean is it's the third time I get up, the third time I see you sitting on the couch, and the second time you've punched me in the face.
The hero *getting up*: - How would I know? You're pissing me off.

The hero punches the heroine who falls K.O. on the ground.

Sequence 7
Sleeping in bed the heroine opens her eyes massaging her face

Sequence 8
The heroine comes in the living room but stays away from the hero.

The heroine: - Tell me... Why did you hit me?
The hero: - I don't know. Because that's the story.
The heroine: - The story? What story?
The hero: - Well... The story of the movie.
The heroine: - What movie?
The hero: - The movie where we're the characters! The main characters!
The heroine: - Ok... But why does the story keeps on starting over and over?
The hero: - I'm not writing it. You should ask the director.

Sequence 9
The heroes are standing in the office of the producer who finances the movie they are the main characters.

The producer: - Please have a seat. So, what can I do for you?
The heroine: - Well it's simple, we would like to meet the director of the movie.
The producer: - Which movie? As you know, I produce a lot of movies...
The heroine: - The movie where we are the main characters.
The hero: - Well, her and me!
The producer: - That part I understood. I'm not a meat head.
The heroine: - It's morning, I wake up. I find this stranger sitting on my couch. I walk up to him kick him out, and he punches me in the face.
The producer *catching a file*: - Ah, that movie! Wait... It doesn't look promising... The director left for the Bahamas to relax, because he has...

The producer waves his fingers.

The hero: - A piano?
The producer: - No, not at all. In fact, he has writer's block.
The heroine: - Writer's block?
The producer: - Yes. He has no idea how to carry on with the story. He has the opening scene: you discovering this guy who never met sitting on your couch, but, for the moment, the story ends right there. So he keeps on rewriting the same part over and over, without knowing how he wants to go on the story, idiot.
The heroine: - So what's going to happen to us?

The producer: - Well, you both are in the shit. I mean… it's annoying. As long as the director doesn't come up with any new ideas, you'll be stuck in the same scene over and over again: the wake-up… and the punch in the face.

The heroine: - So, as long as this idiot doesn't rack his brain, we're stuck?

The producer: - Well, you are the characters in his movie. You're at his disposal. Even more: he has the right of life or death over you.

The hero: - But if you stopped financing him, he would give up his movie?

The producer: - True. But, even if it isn't going to be the blockbuster of the century, the director is kind of my nephew. I'm sure you understand.

The hero *catching the producer by his collar and threatening him with a pen on his throat*: - I'm gonna bleed you like a stuck pig if you don't do anything.

The producer: - Ok, ok, ok, ok. Ok. You won. I'll stop financing the movie!

The producer tears up the file.

Sequence 10

Suddenly the heroes stand in a big white empty room and looks around them.

The heroine: - Fuck! Where are we?

The hero: - No idea. On a blank page, I guess.

The heroine: - Great, what's going happen to us? Did you really have to resort to violence?

The hero: - You don't get it? We are free! Free to do what we want, free to write our own story, free to go wherever we want!

The hero goes to a door standing in the middle of the room.

The hero: - Just imagine a place you'd like to go! So where shall I take you?

The hero opens the door. Behind it there are toilets.

The heroine: - Sorry, I didn't go all day!

The hero *closing the door*: - But not now. You can go later. I don't know. Do you like the countryside?

The hero opens the door. Behind it there are cows in a meadow.

The hero *closing the door*: - Do you prefer the city?

The hero opens the door. Behind it there is the Eiffel Tower.

The hero *closing the door*: -You see. It's simple. You just have to want it. Please, be my guest.

The hero opens the door. Behind it there is a beach.

Sequence 11

The heroes walk on the beach. Then they sit in deckchairs and drink cocktails.

The hero: - So? Isn't this nice?

The heroine: - Yeah. But you've got to like the sea.
The hero: - To like the sea, to like the sea... Isn't it better than taking my punches in your face?
The heroine: - Yeah, I suppose so...

Sequence 12

The heroes are looking at the sea. A director is walking on the beach to them.

The director: - Hello.
The hero: - Hello.
The director: - I'm a movie director and I came up with an idea when I saw you. Would you be interested in acting in a movie?
The hero: - Yeah.
The heroine: - Why not?

Sequence 13

The frame is black. There is only the sound.

The director: - Action!
The heroine: - Hello.
The hero: - Hello.
The heroine: - Who are you? What are doing in my apartment?
The hero: - Ah, so this your place?

The end.

Credits

Cast:

The heroine: Charlotte Normand
The hero: Edouard Blondeau
The producer: Francisco E Cunha
The director: Erwan Orain

Director, author, producer: Edouard Blondeau
Cinematographer: Tristan Chenais
Editor: Romain Prot
Composer: Vincent Bréant
Sound designer: Jocelyn Staderoli

From Brigitte Bardot to Angelina Jolie

Michele Emmer

It is the 1960s, and Brigitte Bardot is at the height of her success. She appears as herself in a film entitled simply *Dear Brigitte* [1]. By simply calling the film *that* everyone immediately knew what is was about. She was the dream girl of the son of the character played by James Stewart. Stewart's character, Robert Leaf, is a poet and professor of English literature in an American university. He is perennially in conflict with the *scientists* at his university, holding science to be dry, and scientific training, particularly mathematics, to be of little use. One day tragedy strikes at home (a sort of tragedy, of course, since the film is a family comedy). Leaf's son, who goes to elementary school, is a mathematical genius. Or better, the boy has a gift for mental calculation. In ordinary language, and thus also in movies, we often find 'mathematical genius' used for those who can perform rapid calculations in their heads. His teacher discovered this by chance, and quite pleased, goes to tell the boy's parents. When he hears that his son is a mathematics whiz (at elementary mathematics, obviously), the father becomes white as a sheet and puts a hand on the mother's shoulder to comfort her. Then, when the teacher leaves, he begins to talk to the boy, begging him not to tell anyone about this skill he has, since it will lead to no end of grief, especially when people begin to shout, when they pass him on the street, 'He is a mathematician!', a phrase Stewart pronounces with disgust, saying, 'And we don't want that, do we?', 'No, Sir', 'Of course not.' Anyway, when the boy goes for a check-up with the doctor, when asked what he is interested in, numbers perhaps, he replies without hesitation, 'Brigitte Bardot'. And when at the end of movie, his dream comes true and he finally sees *la Bardot* in person, he is completely dumbstruck, frozen and blushing in front of his idol. So much for numbers! The movie was made in 1965.

Almost fifty years have passed, and mathematicians have begun to play an important role in films. Their characters are interesting, sometimes even fascinating.

Our heroine, Angelina Jolie, is a British secret agent, actually a double agent, and lover of Alexander Pearce, who cheated the British tax system out of hundreds of millions of pounds. Jolie is continually followed, taped and spied upon. She receives

Michele Emmer
Department of Mathematics, Sapienza University of Rome (Italy).

Emmer M. (Ed.): Imagine Math. Between Culture and Mathematics
DOI 10.1007/978-88-470-2427-4_24, © Springer-Verlag Italia 2012

a message telling her to go to Venice, and in the message, extremely astute (in its way), it is suggested that she pick up a stranger in a train that looks like Pearce.

Now we are on the train, and Jolie is looking for the right person, and she finds, in second class, a man who is smoking. In reality, he has an electronic cigarette that steams and has a light that looks like the burning tip of the cigarette. This is the character played by Johnny Depp. And who would this fortunate man be, chosen to be seduced by Jolie, the man who has to be interesting, resourceful, and fascinating? When she asked him what he does, he answers, 'Math. I teach Math'. And she says, 'I would have not guessed that, I imagine you are a cool math teacher'. 'Still a math teacher', he replies.

The math teacher plays detective, earning right away from Jolie the label of paranoid professor. But actually he isn't that at all. He is nice, kind, helpful, as we gradually discover.

And finally, in the luxurious hotel room where Jolie has invited him, the math teacher kisses her passionately, but maybe it's only a dream...

When the trouble begins – chase scenes, shoot-outs – we see that Depp's mathematican is actually quite agile as he escapes across the rooftops of Venice. And he can dance really well too. Then we learn that it is all a cover, that he is a fake math teacher. It's really Pearce, which was clear from the first close-up of Depp. Right, because the film is predictable, to put it nicely. No suspense, slow motion action, a nothing film. But, the fact remains that the great gentleman trickster Pearce, who has to at least apparently woo Angeline Jolie, chooses to pretend to be a mathematics teacher. This is quite a leap from when James Stewart tells his son that he doesn't want anyone on the street to mistake him in mathematician!

In short, fifty years have passed between the two films, and they haven't passed in vain. In those fifty years it has become credible that a mathematician is not the usual oddball who can't manage his own life, but is so fascinating that he can make Angelina Jolie fall in love with him. We could nitpick and object that in fact Depp's character is not a mathematics teacher, but is just hiding behind that identity in order to appear innocuous and inoffensive, but that is really being picky. The fact remains that they thought of a mathematician. Who knows what the future holds. The film we have been talking about is *The Tourist* [2] directed by Florian Henckel von Donnersmarck, the brilliant director of *The Lives of Others*. Unrecognisable.

In contrast, a very interesting film is *Incendies* by Denis Villeneuve [3], in which one of the main characters is a young mathematician. The film begins with a conversation between her, Jeanne Marwan, and Coen, a mathematics professor at the university where she is an assistant:

Coen: 'Mathematics, as you have known it until today, has tried to provide answers that are certain. Now you are about to enter into an entirely different adventure. The subject will be intractable problems that will always lead to other problems just as intractable. People around you will repeatedly insist that what you are doing is hopeless. You'll have no argument to defend yourself, because the arguments themselves will be of an overwhelming complexity. Welcome to pure mathematics, the land of loneliness. Please meet my assistant, Miss Jeanne Marwan'.

Jeanne: 'Hello'.

We will begin with the *Syracuse Problem* [4]. Film critic Edoardo Becattini compared the film to the proof of a theorem:

> Denis Villeneuve shows himself to be a director with two obsessions: mathematics, and contemporary tragedies. ... *Incendies* is a film constructed like a formula, and the first scene is its equation ... Jeanne's investigation and the life of 'Mother Courage' Niwal represent in fact proof and corollary of the same statement: two paths that not only arrive at the same truth, but also narrate, in essence, the same story twice. But redundancy doesn't frighten Villeneuve. He knows that the mathematician only creates certainties and thus avoids leaving any possible doubt, building the tension by recurring to a logic that is so ironclad that he thinks he can make even the most paradoxical numeric expressions $(1+1=1)$ credible. The film's ambitions are thus many and lofty, but the twists of the story and of contemporary politics, so thorny and undecipherable, are not well-suited to the smooth perfection of the mathematical functions (quoted in http://mymovies.it). I think that the director fully understood the difficulty of telling such a complex story about war, fanaticism and hate in a way that was logical and consequential. The film takes place in Lebanon during one of the many civil wars between Muslims and Christians. With the opening scene, Villeneuve tells us that being a person who is rigorous, intelligent, consequential – in short, being a mathematician – as the paradigm of the capacity to face new, unpredictable and complicated situations, is precisely the trait which is needed. Without having to give up one's own feelings, fears, and inadequacy.

Mathematician Coen says, 'What does your intuition tell you? Your intuition is always right. This is why you have what it takes to become a real mathematician'.

As another critic, Fabio Ferzetti, has underlined in a review in *Il Messaggero* (January 21, 2011), the director insists on 'an intellectual framework – Jeanne is a talented mathematician – which makes that ungovernable chaos even more cruel ... We are on very high ground, capable of uniting blood and abstraction, the tumult of bodies and the incessant intrigue of intelligence and of pity'.

Didn't Plato say that perhaps the world should be governed by mathematicians?

> Geometry is the knowledge of the eternally existent [...] [geometry] will tend to draw the soul to truth, and would be productive of the philosophical attitude of mind, directing upward the faculties that now wrongly are turned earthward. [...] Then nothing is surer than [...] than that we must require that the men of your fair city shall never neglect geometry [5].

Fig. 1 *Incendies* by Denis Villeneuve

In *Scorched* by Wajdi Mouawad, the play on which the film was based, the connection with mathematics is made even more explicit. The lesson that Jeanne Marwan gives is on graph theory. In the course introduction that she gives to the students, she says:

> People will often criticise you for squandering your intelligence on absurd theoretical exercises, rather than devoting it to research for a cure for AIDS or a new cancer treatment. You won't be able to argue in your defense, since your arguments themselves will be of an absolutely exhausting theoretical complexity. Welcome to pure mathematics, in other words, to the world of solitude. Introduction to graph theory [6: p. 11].

Jeanne tries to use the structure of graph theory to understand what is happening, establishing a connection between the polygons of the theory and the people that she has to deal with:

> Beginning with a theoretical application, for instance, can I draw the visibility graph and the corresponding diagram? What is the shape of the house where the members of the family represented in the application live? Try to draw the polygon. You can't do it. The whole of graph theory rests essentially on this problem, impossible to solve for no. Now, it is this impossibility that is wonderful [6: p. 13].

Faced with the devastating information that the notary gives her following the death of her mother, she cries, 'What is my place in this polygon? My father is dead, that is the conjecture. Everything leads me to believe that it is true. But nothing proves it. I haven't seen his body, nor his grave. So it could be that, between 1 and ?, my father is alive'. The young mathematician often tries to construct a graph to explain what is happening to her. It will be her mother who tells her that her place is in the very heart of the polygon.

In 2010, thanks to the courtesy of Gianni Amelio, I had the chance to see his first film, created for RAI in 1979 and absolutely impossible to find [7]. This too is the story of a child prodigy who is interested in mathematics and music. Amelio's film is based on a story by Aldous Huxley, *Young Archimedes,* published for the first time in 1924 [8]. When Clifton Fadiman published the anthology of stories related to mathematics in 1958, he put Huxley's story as the first one. The main character is an English scholar of medieval Italian painting, Mr Heinz. The film is slow, contemplative, as gentle as the Tuscan hills where it takes place, and in which nature also becomes a kind of character. Into this kind of Eden comes a boy, Guido, the son of farmers, without any education at all. Mr Heinz's son, the same age, is captivated by his vivaciousness and curiosity. Guido discovers music, hears some records. He immediately recognises in the music he is listening to two violins that play different melodic lines, and gradually he discovers and falls in love with Mozart, Bach. He begins to play the piano, he is a natural talent, and after a short time, he begins to write music.

The film's bad guy, or better, the bad gal, is the mistress of the house where Mr Heinz lives, Mrs Bondi, played by a mysterious Laura Betti. She doesn't have any children, and she wants to adopt the little farm boy, make him her own.

Fig. 2 *Fantasia Mathematica* edited by C. Fadiman, 1958

In the meantime, Mr Heinz discovers another extraordinary quality in the boy:

What I actually saw was Guido, with a burnt stick in his hand, demonstrating on the smooth paving stones of the path, that the square on the hypotenuse of a right-angled triangle is equal to the sum of the squares on the other two sides. Kneeling on the floor, he was drawing with the point of his black-ended stick of the flagstones ... and he proceeded to prove the theorem of Pythagoras – not in Euclid's way, but by the simpler and more satisfying method which was, in all probability, employed by Pythagoras himself... In a very untechnical language, but clearly and with a relentless logic, Guido expounded his proof ...

Guido: 'Wait a moment. But do just look at this. Do. ... It's so beautiful. It's so easy.'

So easy ... The theorem of Pythagoras seemed to explain for me Guido's musical predilections. It was not an infant Mozart we had been cherishing; it was a little Archimedes with, like most of his kind, an incidental musical twist ...

'Guido, who taught you to draw those squares?' 'Nobody.' 'You see, it seemed to me so beautiful ...'

Mr Heinz concludes that, 'Perhaps the men of genius are the only true men. In all the history of the race there have been only a few thousand real men. And the rest of us – what are we? Teachable animals'. The story takes place during the Fascist era.

Guido has never been to school, but he can read and write. Mr. Heinz begins to teach him formal mathematics. At a certain point the idyll is broken. Heinz has to leave, staying away for several months. He receives a letter from Guido, who has learned to write a few words. It is a cry for help: the mistress has taken away all the mathematics books, he can't play the piano anymore, he doesn't like how things are, and he asks for help.

Heinz comes straight back, but the boy is dead, either by purposely throwing himself out of a window, or by accidentally falling. This is the price that the simple but brilliant boy pays for being different, for being exceptional.

Final observations

Mathematics is not rigidly applied in films like it is in art, because here mathematics – as the last twenty years of films and theatre inspired by mathematics, or better, by mathematicians, have shown – is used for its power to inspire, evoke, mystify, and fascinate. As Max Bill said about the relationships between modern art and mathematics:

> By a mathematical approach to art it is hardly necessary to say I do not mean any fanciful ideas for turning out art by some ingenious system of ready reckoning with the aid of mathematical formulas. [...] It must not be supposed that an art based on the principles of mathematics [...] is in any sense the same thing as a plastic or pictorial interpretation of the latter. Indeed it employs virtually none of the resources implicit in the term 'pure mathematics'. The art in question can best be defined as the building up of significant patterns from the ever-changing relations, rhythms and proportions of abstract forms, each one of which, having its own causality, is tantamount to a law unto itself. As such, it presents some analogy to mathematics itself [9: p. 5, 8].

References

1. *Dear Brigitte*, directed by H. Koster, with J. Stewart, B. Mumy, G. Johns, B. Bardot, story and screenplay by J. Haase, N. Johnson and H. Kanter, USA, 1965.
2. *The Tourist*, directed by F. Henckel von Donnersmarck, with J. Depp, A. Jolie, P. Bettany, screenplay by F. Henckel von Donnersmarck, C. McQuarrie and J. Fellowes, USA, 2010.
3. *Incendies*, directed by D. Villeneuve, with L. Azabal, M. Déormeaux-Poulin, M. Gaudett and R. Girard, screenplay by D. Villeneuve, with V. Beaugrand-Champagne, based on a play by W. Mouawad, Canada, 2010.
4. Also known as the *Collatz Conjecture* after the mathematician Lothar Collatz who proposed it in 1937, and by other names as well, the conjecture regards whole numbers. Take any natural number n, that is, a number like 1, 2, 3, 4, and so forth. If n is equal, divide by 2; if it is odd, multiply by 3 and add 1. If this process is continued, then no matter what number you start out with, you always end up with 1. The conjecture has never been proven.
5. Plato, The Republic, Book VII: 527b-c (trans. Paul Shorey).
6. W. Mouawad, Scorched. Playwrights Canada Press, 2005.
7. *Il piccolo Archimede*, directed by G. Amelio, with L. Betti, J. Steiner, A. Salvi, F. Pugi, story by A. Huxley, screenplay by G. Amelio, photography by G. Bertoni, music by R. Vlad, production RAI, Italy 1979.
8. A. Huxley, Young Archimedes (1924). In: *Fantasìa Mathematica*, Clifton Fadiman, ed. Simon and Schuster, New York, 1958, pp. 3-35.
9. M. Bill, 'A Mathematical Approach to Art' (1949) reprinted with the author's corrections. In: M. Emmer (ed.), The Visual Mind: Art and Mathematics, MIT Press, Boston, 1993, pp. 5-10.
10. M. Emmer, Numeri Immaginari, cinema e matematica. Bollati Boringhieri, Torino, 2011.

Homage to Luca Pacioli

The Mathematical Ideas of Luca Pacioli Depicted by Iacopo de' Barbari in the *Doppio ritratto*

Enrico Gamba

Of the paintings that are part of Urbino's history, the one that is usually considered the most mysterious and problematic is Piero della Francesca's *Flagellation*. But there is another painting that also comes from Urbino which is just as mysterious and problematic: the so-called *Doppio ritratto*, or dual portrait, by the Venetian painter Iacopo de' Barbari.

In its own way, each painting embodies the definition of *mathematical humanism*, the expression coined by André Chastel to describe the cultural climate of the 1400s, highlighting the peculiar and distinctive role played by mathematics in the cultural environment of the Urbino court with respect to other courts in Italy. Mathematics, in the sense of the term of those times, comprised the traditional arts of the quadrivium – arithmetic, geometry, astronomy-astrology, and music – and extended to perspective, architecture, and the construction of machines for civil and military use.

Going back to the paintings mentioned, we should note right away that an enormous number of specialised studies have been written about the *Flagellation*, including books conceived for a popular readership of non-specialists, while the *Doppio ritratto*, in spite of its being well-known and frequently reproduced, hasn't been the object a similar share of attention by critics.

The aim of this paper is to examine only the painting's mathematical aspects, the typical fruit, or better, visual expression, of Urbino's *mathematical humanism*. I will not, for instance, enter into the question of the identity of the person to whom Luca Pacioli is giving a lesson in mathematics. I will only say that the erudite Urbino mathematician Bernardino Baldi reported in 1589 that the painting was kept *ne la guardaroba de' nostri serenissimi Principi in Urbino*, 'in the wardrobe of our most serene Princes in Urbino' (a wardrobe in the Renaissance was more than a closet; it housed precious belongings).

From the point of view of the history of mathematics, this painting by Iacopo de' Barbari is the first portrait of a living, working mathematician, surrounded by the *tools of the trade*, caught *in the act*, so to speak. It is not an imaginary portrait of the *auctoritates* such as Euclid, Archimedes and Ptolemy, nor are the figures accompanied by stereotypical objects in the way, for example, Ptolemy is often portrayed

Enrico Gamba
Centro Internazionale di Studi "Urbino e la Prospettiva", Urbino (Italy).

Emmer M. (Ed.): Imagine Math. Between Culture and Mathematics
DOI 10.1007/978-88-470-2427-4_25, © Springer-Verlag Italia 2012

Fig. 1 Iacopo de' Barbari, *Doppio ritratto*, Museo e Galleria Nazionali di Capodimonte, Napoli

next to an astrolabe. To the contrary, this painting presents a precise narrative: Pacioli has personally guided the hand of Iacopo de' Barbari, indicating down to the smallest detail what to represent and how to do it; everything in the painting is mathematically purposeful and contrived. The aim is to communicate visually the concept of mathematics professed by the Franciscan friar; if you will, the painting functions as a kind of visual curriculum vitae for Pacioli.

In short, Pacioli's aim was to present a global image of mathematics, a discipline that came to include three components: the rational-scientific, the mystical-philosophical, and the technical-practical. These three components – or better, traditions – are clearly and knowingly illustrated in the painting. Let us look at them in order.

On the edge of the frame of the small slate on the left of the table is written EUCLIDES, while the open book on which Pacioli's left hand rests is a copy of the first edition of Euclid's *Elements*, published in Venice in 1482. The text is depicted with absolute precision, allowing us to see easily that it is opened to proposition 8 of Book XIII.[1] It hardly needs saying that the *Elements*, for centuries and centuries, was the text responsible for the spread of a rigorously deductive mathematics, whose propositions enjoyed absolute certainty.

---

[1] Before the Greek text of Euclid's *Elements* prepared by Simon Grynaeus (1493-1541) in 1533 (and occasionally afterwards as well, as shown by the Italian translation by Tartaglia), the *Elements* were known only in Latin translations of Arabic translations from the Greek. In all these translations, the proposition in Book XIII which in Greek we find denoted as proposition 12 is indicated as proposition 8. In the portrait of Pacioli, the book on which his hand rests is his own edition of 1509, which he says is the corrected text of the 'translation' (in quotation marks because it the interventions are major) done by Campano in the mid-thirteenth century, based on a variety of Arabic sources. Thus, although the modern reader will identify it as proposition 12, here it is identified as proposition 8 because we are referring to the text that Pacioli knew. Many thanks to Fabio Acerbi, editor of *Euclide, Tutte le Opere*, Milan: Bompiani, 2007.

The philosophical-mystical tradition is not represented by a book such as, for example, the *Arithmetica* by Boethius, but rather by two geometrical figures: a dodecahedron resting on a copy of Pacioli's *Summa de arithmetica, geometria proportioni et proportionalità* printed in Venice in 1494, and a rhombicuboctahedron.

The dodecahedron is made out of wood, and is intentionally an object that is material, physical and concrete. In contrast, the rhombicuboctahedron is a metaphysical object: it is a Platonic idea that appears before our eyes; an idea that comes from on high, because there is no way to tell what it is hanging from; an apparition that suddenly materialises out of the black background. It is, in fact, an object that, technically speaking, cannot be built; it appears to be made of the perfect crystalline material that the celestial spheres were believed to be made out of. This is a declaration in favour of the Platonic nature of mathematical objects. Mathematical objects are part of an ideal, perfect world, a reality of a higher order, certain, immutable; this was Pacioli's belief. However, they are not separate, unattainable entities: there is the crystalline, ideal polyhedron and there is its material equivalent, the wooden polyhedron. Pacioli was a man of the world, not a metaphysicist. He uses Platonism to affirm the cultural dignity of mathematics within the cultural debates of the day, a dignity that he believes to be valid for all of mathematics, commercial and technical included. Here Plato would disagree. The wooden dodecahedron provides assurance that the pure mathematical ideas can be applied to the concrete things of this world; this is an advantage, not a outrage.

Now we arrive to the technical-practical component, fittingly represented by the conspicuous volume of the *Summa*, which, although closed, Pacioli wants to make sure is recognisable by the writing on the edges: L[IBER] R[EVERENDI] LUC[AE] BUR[GENSIS]. The *Summa* was dedicated to the Duke of Urbino Guidubaldo da Montefeltro, and remained widely-read and influential throughout the 1500s. In its first pages we find examples of numeric philosophy. It is sufficient to cite the example of the qualities of the number 1:

Essa unità non è numero, ma ben principio de ciascun numero, ed è quella mediante la quale ogni cosa è detta essere una. E secondo el Severin Boezio in sua musica, è l'unità ciascun numero in potentia.

(This unit is not a number, but rather the beginning of each number, and is that by means of which each thing is said to be one. And according to Boethius in his music, the unit is each number in potential).

However, after these first few pages the general structure of the work is clearly practical. Outstanding among problems of capitalisation, society, measuring of areas of land, and so forth, is the treatise *De computis et scripturis*, the first printed version of the techniques of commercial bookkeeping. The famous double-entry bookkeeping had already been used by merchants for a couple of centuries; Pacioli turned it into an instrument for understanding and managing the economic, financial and patrimonial aspects of commercial activities.

Finally, the painting poses a challenge, something that was not at all unusual among mathematicians of the day. Drawn on the small slate mentioned earlier, is a circle in which is inscribed an equilateral triangle, from whose upper vertex is drawn a line that is purposely left incomplete. With one hand Pacioli points to proposition

8 of Book XIII of the *Elements*; in his other hand he holds a stick that points to the incomplete line. All that is missing is his voice saying, 'And now how do we go on? How do we put these things together?'

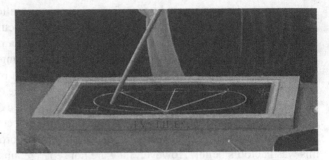

Fig. 2 Detail of the Euclides' slate

One possible answer is that the line will end at point X, so that the completed figure will be a square inscribed in the circle. According to proposition 8 of Book XIII, the square of the side of the triangle is three times the square of the radius; it states the relation between the radius of the circle and the side of the triangle inscribed in it.

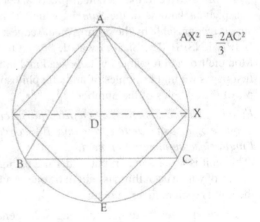

$$AX^2 = \frac{2AC^2}{3}$$

Fig. 3 Possible completion of the figure drawn on Euclides' slate

The immediate next step is to derive the relation between the radius and the side of the square inscribed in the circle. From this is easily derived the relation between the triangle and the square: if the square of the side of the square is equal to twice the square of the radius, then the square of the side of the square is 2/3 of the square of the side of the equilateral triangle.

In my opinion, this hypothesis finds an indirect confirmation from the exact numeric calculation that appears in the lower right-hand corner of the slate: 478 + 935

+ 621 = 2034. How might these numbers be interpreted? Were they chosen at random? Knowing Pacioli, it is hard to believe so. At first glance the numbers do not appear to be special; for instance, they are not perfect numbers, or amicable numbers, or Fibonacci numbers, etc. Note that no digits are repeated; this might lead us to think of a magic square arranged such that the sums of the numbers along the rows, columns and diagonals are constant, but this is not the case. The only harmony, if you will, that I have found is between 621 and 478, that is, if 621 is the perimeter of an equilateral triangle with a side of 207, then 478 is the perimeter of a square constructed on the radius of the circle in which that triangle is inscribed. In other words, in proposition 8 of Book XIII, Euclid the geometer puts the surfaces into relation, while Pacioli, who is also an arithmetician and algebraist, puts specific cases of the perimeter into numerical relation, which is the customary way to treat specific numeric cases in the context of practical mathematics, that is, abacus mathematics.

In conclusion I would like to remark that the characteristic that renders the painting absolutely unique is that it unites a cutting-edge work of art – that is, the representation of an empty, transparent rhombicuboctahedron – with cutting-edge avant-garde science – such was the mathematics of polyhedra at the time. I believe that no other painting in the whole history of art has been able to do that, thus providing us with a unique example of a close collaboration between a mathematician and a painter in order to achieve the depiction of the avant-garde in both mathematics and painting.

References

1. A. Ciocci, Luca Pacioli tra Piero della Francesca e Leonardo. Aboca Museum Edizioni, Sansepolcro, 2009.
2. Euclid, Euclide. Tutte le Opere. Introduction, translation, notes and critical apparatus by Fabio Acerbi. Bompiani, Milano, 2007.
3. E. Giusti (ed.), Luca Pacioli e la matematica del Rinascimento (Conference proceedings). Petruzzi, Città di Castello, 1998.
4. Descriptions of the "Doppio ritratto" by M.G. Ciardi Dupré and M. Seracini. In: Piero e Urbino. Piero e le corti rinascimentali (exhibition catalogue), edited by P. Dal Poggetto, Marsilio, Venezia, 1992, pp. 200, 466-468.

De divina proportione: from a Renaissance Treatise to a Multimedia Work for Theatre

Simone Sorini

The theatrical work *De Divina Proportione*, which made its debut in Urbino in 2009 – the year which celebrated the five-hundredth anniversary of the print publication of the book by Luca Pacioli – and performed again with success during the Ravenna Festival of 2011, is a multimedia spectacle with live music, dance and video projections inspired by – indeed, taken from – the famous text of 1509. Pacioli's text is not only considered to be a watershed of scientific knowledge of the day, but one which in many respects is interdisciplinary. It was precisely in virtue of the work's interdisciplinary nature that we developed the idea of bringing an ancient mathematical text to the theatre in the form of a musical work in a modern key. In the very first pages of the treatise we come upon an interesting declaration: the author explains at the outset that his study will be useful and necessary for all 'perspicacious and curious minds' interested in philosophy, perspective, painting, sculpture, architecture, music and other kinds of mathematics.[1] Where do we begin, what should we take from a five-hundred year old treatise to construct a musical work for the present day?

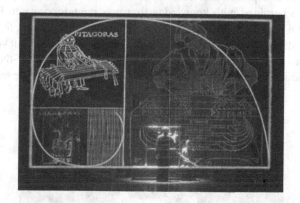

Simone Sorini
Bella Gerit, Urbino (Italy).

[1] *Opera a tutti gli ingegni perspicaci e curiosi, ove ciascun studioso di Philosophia, Perspectiva, Pictura, Sculptura, Architectura, Musica e altre Matematiche, suavissima, subtile e admirabile doctrina conseguirà e delectarassi con varie questione de secretissima Scientia'* [Luca Pacioli, *De Divina Proportione*, Venezia, 1509].

Emmer M. (Ed.): Imagine Math. Between Culture and Mathematics
DOI 10.1007/978-88-470-2427-4_26, © Springer-Verlag Italia 2012

First of all, the choice on the part of musicians like us to concern themselves with things of mathematics might seem at first glance to be ill considered. But this isn't so if we think that from the times of antiquity music was placed alongside her sister arts, arithmetic, astronomy and geometry, to make up the *quadrivium*, which formed the basis of all scientific studies. Another factor that played a significant role in this choice was our being from Urbino. The term 'mathematical humanism' was coined for the city of Urbino, because of the distinct role played by the city during the Renaissance with regard to the sciences and mathematics. You only need to wander the streets and by-ways of the city to see that among the many plaques commemorating the birth of illustrious figures, a large number regard mathematicians, builders of compasses and precision instruments, military architects, and so forth. Moreover, a significant percentage of the hundreds of volumes housed in the famous library of Federico da Montefeltro comprised works both ancient and contemporary of a scientific and technical nature, a much higher percentage than other famous libraries of the day. Urbino can boast of a genuine vocation for the scientific disciplines, and Pacioli went there on several occasions to study and work in the court of the Dukes, where he had at his disposal all of the precious library's instruments. Without entering into the question of the famous accusation that Pacioli had plagiarised the work of Piero della Francesca, first formulated by Vasari and seconded by many others (Piero had written a treatise on the Platonic solids for Federico, which was kept in the library just mentioned, and Pacioli included it in the final printed version of his treatise without citing the source, in spite of the fact that he had carefully noted the sources he used for the classics), Pacioli surely deserves credit for being able to create a work that is *beautiful* as well as scientific, with the kind of farsightedness that characterizes a good editor today: of knowing how to add scientific experience and artistic skills together, in other words, to construct a bridge between the two disciplines. Another bridge can be seen in the way ancient scientific knowledge is joined with knowledge that was contemporary to Pacioli's day, so as to create a reliable reference work for scholars. Furthermore, the entire book intended for distribution in print, and was in fact printed in the Italian vulgar and not in Latin, as was the canonical form for this kind of work. Both of these decisions surely corresponded to a desire to disseminate the work among the humanists, but

also to make the work accessible to many thanks to the new technique of printing, which would soon profoundly change the entire cultural panorama.

In the process of creating our theatrical work, we kept uppermost in our minds the fact that for Pacioli himself, the effects of mathematics were often considered to be spectacular, supernatural, a kind of practical illusion capable of arousing wonder and amazement in those present.

Pacioli never underestimated the power of these instruments in carrying out his intention of spreading knowledge, and to this end he wrote at least two treatises, the *De ludo scachorum* and the *Viribus quantitatis*, written a few years before *De divina proportione*.

As to the importance that the author attributed to the arts, we know that Pacioli was constantly in the company of the artists of the day, especially the painters he himself defined as 'geometric painters', many of whom he knew personally, including – in addition to Leonardo (who drew the polyhedra for *De divina proportione*) Sandro Botticelli, Domenico Ghirlandaio, Piero de la Francesca and Albrecht Dürer. These painters used perspective as a genuine mathematical science. We know that the study and application of mathematics in the 1400s were disseminated in at least three different areas: in university teaching, in that of the abacus schools, and in the *botteghe*, or workshops, of the artists, where knowledge of at least elementary notions of abacus mathematics was mandatory in order to correctly set up what was considered by the artists to be the discovery of the century, that is, perspective.

In this regard there is a dispute that Pacioli raises in the third chapter of the treatise: he proposes adding perspective to the four arts of the *quadrivium*, and if this is found objectionable, then he proposed taking away music, because in the same way that music delights the ear, painting delights the eye, and so forth. This is clear evidence of Pacioli's complicity with the painters of his day, who shared his ideas not least because the painters at that particular period felt a strong need to redeem themselves from the *mechanicalness* of the profession of the painter, which was not considered to be far removed from the humble labour of the blacksmith, which was also learned after a long apprenticeship in the workshop. The aspirations of the *geometric* painters in Pacioli's circle – we need only think of Leonardo – were of a completely different nature: to all intents and purposes, they were scientists.

The theories of proportion and perspective expressly refer to Euclid; the so-called visual pyramid is drawn from his treatise on optics.

Proportions and anthropometry thus represent a paradigm, one capable of uniting painting to architecture, and architecture to music. On the other hand, isn't mathematics still today, like art, a universal language.

Given these introductory remarks, what constitutes the particular nature and originality of the theatre piece? Without a doubt, it is based on choices that are *rational* as well as aesthetic. Everything that is heard, seen and perceived in the theatre is the result of numerical applications, completely in keeping with Pacioli's treatise. The design of the performance itself respects the three-part division of Pacioli's book of 1509; our project is divided into three distinct sections in golden ratio with respect to each other, cadenced by voices off-stage (those of mathematician Piergiorgio Odifreddi and actress Lucia Ferrati), who read several passages of the original text by Pacioli, thus making it possible to listen to the sound of this beautiful antique Italian technical language.

The first part is that in which Pacioli states the thirteen admirable effects of the Divine Proportion (thirteen, like the number of Christ added to that of the apostles), after having presented the characteristics and properties of the segment divided according the *mean and extreme ratio*, he chooses four that are similar to the divine attributes: the first is uniqueness, the second is the trinity, the third is the irrationality, the fourth is immutability. There is also a fifth characteristic which is above all others in importance: just as the nature of God is identified with heavenly Virtue, also called the Fifth Essence, or *quintessence*, and infuses all of the other four simple solids, that is, the four elements Earth, Air Fire and Water, in the same way, by means of this proportion to heaven is attributed the figure of the dodecahedron, or solid of twelve pentagons – in accordance with Plato's *Timaeus* – whose construction is based on the divine proportion. In an analogous way, a form is assigned to each of the other elements: to fire the pyramidal solid called the tetrahedron; to the earth the cubic solid called the hexahedron; to air the solid called the octahedron; and to water the icosahedron. These solids are defined as the regular polyhedra, and from them it is possible to obtain the numerous variations that Pacioli calls *elevate* (star polyhedra) or *abscisse* (truncated polyhedra). According to Pacioli, everything arises from the golden ratio, which is the building block of the five polyhedra; on these are based all that can be called material.

The golden section is represented in our theatrical work by the monochord, a Pythagorean instrument which has a single chord that can be divided by means of a moveable bridge, which reproduced it symbolically (even though in reality the division of the chord to produce intervals that are pleasing to our ears and recognisable is fixed according to precise ratios that do not have anything to do with the golden section).

The entire universe of sound in the theatrical work originates in the physical sounds of the monochord. These sounds, comprising the intervals of a fourth, a fifth, a major third and the golden section, are recreated by electronic means and spatialised in a three-dimensional acoustic context by means of original software created expressly for this purpose.

The machines interpret these intervals, elaborate them and give them back by means of the ambisonic equipment in the form of a cube installed in the theatre, in virtual form. In effect, all of the first part, that dedicated to the golden section and the constitution and inscription of the polyhedra, is centred on the metaphysical world. The Platonic world of ideas is also represented by the constant presence of a scrim on stage, onto which are projected the images and videos created by Pierluigi Alessandrini.

Following the finale of the first part, through the construction of the sphere following the intersection of the polyhedra with one another, we arrive at the second part, where the harmonic series is constructed acoustically, and pure sounds are left behind.

The second part is dedicated to the elements and their correspondence to the polyhedra, in the way this was illustrated in antiquity by the Pythagorean School, and in particular by Empedocles of Agrigentum. At this point in the performance we found ourselves immersed in a harmonic acoustic universe, created by physical instruments (since antiquity, each instrument has been associated with an element), which improvise on the theme of the elements, against an electroacoustic background of sound landscapes. The final element to be represented is the quintessence, which we interpret entirely with electronic means.

The final, third part is that inspired by the treatise on architecture contained in the print edition of *De divina Proportione*. The screem is removed from the stage, and all of the musicians and dancers appear, without the concealing diaphragm of the projections. We are now in the physical world, and man projects his measurements and his proportions onto works of architecture. The entire section is permeated by the concept of anthropometry. With regard to the choice of ancient works of music, we were extremely stimulated by the clear conviction that we had to start with numbers, and only with numbers, to create a universe of sound; obviously, we would also have to take into account what history had produced, and there were no lack of examples: Guillaume de Machaut, Philippe De Vitry, Guillaume Dufay. All of these masters of the late Middle Ages – although in the case of Dufay we begin to speak,

not entirely correctly, of the Renaissance – based their works, especially their sacred works, on numerical allegories, isorhythms, canons, palindromes and the like.

Because of his proximity to Italy and because he is the closest in time to early humanism, it seemed obvious to us that we had to consider Dufay an important element in our musical study. Moreover, this master's perhaps most universally famous work was the celebrated motet *Nuper Rosarum Flores* which he composed and probably directed, *organised* you would say in Italy, specifically in Florence on the occasion of the consecration to the Virgin Mary of the Cathedral of S. Maria del Fiore, on 25 March 1436. This work, as musicology students know quite well, possesses interesting characteristics of structure and composition. In fact, it is an extremely original form of musical hermeneutics, since the whole composition is based on some sacred numbers relative to the Solomon's kingdom and temple, as passed down in the first book of Kings in the Old Testament. Further, according to the thesis by Craig Wright[2] we find a particular and insistent reiteration of the number 7, or 7 x 4 – the number of the Virgin characterized by her seven sorrows, the seven days of exile in Egypt, her seven virgin companions – and the number 4, found in almost every ecclesiastic symbol. The formal framework consists of a *cantus firmus* that the two tenors execute in notes that are long and rhythmically staggered at a distance of a fifth on the motive *Terribilis est locus iste*.

The number 7 is found in the *Nuper* meanwhile in the choice of the seven-syllable lines arranged in verses of seven strophes of the Latin text, perhaps the work of Dufay himself, and thus in the 56 (7x4=28x2) breves in *integer valor*, or whole beats, of each of the four sections of the motet for four voices.

The piece is divided into four parts, each of which comprises an exposition of the *cantus firmus* a different metric indication. This means that the durations are different for each section, even though each contains an identical number of beats (fifty-six, of which the first twenty-eight are intoned only by *motetus* and *triplum* and in the remaining parts they unite the two tenors in the melody of the *introito*).

However, while the number of beats is equal, the *tactus* (which corresponds to the unit of measure of the pulse rate of a man breathing normally) differs from section to

2 C. Wright, Dufay's 'Nuper rosarum flores', King Solomon's Temple, and the Veneration of the Virgin. Journal of the American Musicological Society, vol. 47, no. 3, 1994, pp. 395-427 and 429-441 (46-page article).

section. The values obtained for the *tactus* are 168, 112, 56 and 84, which, divided by 28, give the ratios 6:4:2:3.

In fact, the mensural ratios of the four sections, excluding the brief amen, are in the particular proportion 6-4-2-3. In modern terminology we say, although not entirely correctly, that we go from a tempo of 6 to 2, 4 and 3. If we look at 1 Kings 6:2, we see that according to the measurements given in the description of the temple that King Solomon had built in Jerusalem for his father David, the length was 60 cubits, the width was 20 cubits, and the height was 30 cubits, while the cell destined to contain the ark of the covenant was 20 cubits. All of the musical structure makes clear reference to the temple's architectural dimensions, without counting the fact that musicologist Charles Warren[3] has identified in Brunelleschi's cathedral the same proportions expressed in Florentine *braccia*, that is, once again the module used by Dufay. This means that the musician and the architect worked together, or if you prefer, that the musician built his structure on the basis of a clear dimensional symbolism.

This could not help but bring to mind the treatise on architecture and anthropometry contained in the book by Pacioli. Surely the *Nuper Rosarum Flores* is the most distinguished example of musical architecture of the past, and perhaps of all time. Thus the final part of our performance, that which takes the name *Grandis Templum Machinae*, is dedicated to this composition, which is taken apart and reconstructed by an electronic sound apparatus for electroacoustics, recomposed starting from its *talee* and its *colour* in a virtual, technological context. The new composition is comprised of seven parts (in comparison to four parts in the original). The three additional parts are inserted after the first, second and third parts of the original motet, and are intended to build on the sonorous considerations on the *modus*, *tempus* and the *prolatio* expressed by Dufay through the proportions of the structure, acting as a commentary and introduce the new *integer valor* for the breve, anticipating it with rhythmic solutions obtained from the sound spectra generated by the computer. In effect this is a temple within a temple; our *Grandis Templum Machinae* contains the older composition, just as allegorically the Temple of Solomon contained *in nuce* the entire Christian church.

---

[3] C.W. Warren, Brunelleschi's Dome and Dufay's Motet. In: The Musical Quarterly, vol. 59, no. 1, 1973, pp. 92-105 (14-page article).

Credits

Design and direction Simone Sorini/David Monacchi
Video recording and editing Pierluigi Alessandrini
Sets and lighting coordination Andrea Maria Mazza

Damien Fournier *contemporary dancing*
Clio Gaudenzi *contemporary dancing*
Claudia Viviani *female figure with polyhedra*
Andrea Angeloni and Luigi Germini *trombones*
Matteo Bellotto *bass*
Angelo Bonazzoli countertenor Enea Sorini *baritone e psalterium*
Simone Sorini *ensemble director, tenor, lute*
David Monacchi *sound director, real-time spectrogram, flute*
With special appearances by:
Piergiorgio Odifreddi and Lucia Ferrati *virtual voice recitations*

From an idea by Simone Sorini
Electroacoustic composition David Monacchi

Original software used for electroacoustic sound generating:
Tonharmonium by David Monacchi and Aaron McLeeran, *Stria by* Eugenio Giordani
Sensor management by LEMS - Laboratorio Elettronico per la Musica Sperimentale

Multichannel amplification: Enzo Geminiani - SwanSound

Initial consultants:
Duccio Alessandri *hermetics*
Laerte Sorini *mathematics*
Christian Cassar *scenography*
Giovanni Caffio, Giuseppe Marino (Studio Ippozone) *graphics, visual layouts*
a production by Bella Gerit Urbino

http://www.de-divina-proportione.it

Mathematics and Music

Towards a Rational Practice of Arithmetic. A Model for Musical (and Multimedia) Composition

Francesco Scagliola

1 Introduction

This paper proposes some practices in the field of art music composition. For the type of insights and the approach taken, the theory and practice proposed can only be computational. An introduction to the techniques necessary for the Computer-Aided Algorithm Composition (CAAC) can be found in [5], [1], [10] and [13], with particular reference to *Formalized Music* by Iannis Xenakis [16].

The choice to adopt the computer as a possible instrument available to the composer comes from the idea that appropriate theoretical models may be capable of unravelling the communication discomfort in which some composers find themselves after the experience of the historical avant-garde.

A lot of artistic production of the historical avant-garde was created under the influence of Marxian ideology dominant in the cultural circles of the Second World War. In light of this consideration, it is possible to interpret properly, for example, the work of Pierre Boulez (b. 1925) [3] or Luigi Nono (1924-1990) [14]. It is worth noting that dissenting voices arose against some musical techniques derived from the Marxian ideology, such as that of György Ligeti (1923 -2006) [9]. Nowadays, only the strictly musical compositional techniques developed during the avant-garde age remain as common practice.

Furthermore, during the period of avant-garde predominance different kinds of music were successfully developed, such as those of Benjamin Britten (1913-1976), Leonard Bernstein (1918-1990) and the minimalist Steve Reich (b. 1936).

However, in recent decades composers have opted for composing in ways that depart from dogmas belonging to a bygone era. Most of them compose on the basis of both the musical instinct, which must be the first nature of a composer, and music theory, which appears to be inadequate with respect to the current expressive needs.

Francesco Scagliola
Conservatorio di Musica "Niccolò Piccinni", Bari (Italy),
Founder and Director of **Sin[x]Thésis**, Group of Multimedia Research and Production (Italy).

Emmer M. (Ed.): Imagine Math. Between Culture and Mathematics
DOI 10.1007/978-88-470-2427-4_27, © Springer-Verlag Italia 2012

For these reasons, some composers of the younger generation feel the need for an overhaul of the theory so that it is no longer tied solely to historical and social needs, as happened, for example, in the case of both the Darmstadt *Ferienkurse für Neue Musik*, and philosophical urgency, masterfully expressed by John Cage (1912-1992).

The new theory should be derived from music itself, which is difficult because the music is considered the ineffable art par excellence [7]. In fact, Gottfried Wilhelm Leibniz (1646-1716), wrote in a letter dated 17 April 1712 to Christian Goldbach (1690-1764): Music is a hidden arithmetic exercise of the soul, which counts unconsciously[1].

In this context, the objective of this paper is to show some hypotheses to rationalize this occult arithmetic practice, that is music, to shed light on general issues which can guide the practice of music composition.

Finally, I would like to emphasize that this paper is written from the point of view of a composer, which differs from the prospective of a mathematician, physicist, engineer or computer scientist. As a musician, I am interested in the results of computational models and not in the models themselves.

Indeed, brilliant models not necessarily produce musically meaningful results.

2 Symbols in Music

It is a common practice in music analysis to represent the entire piece or parts of it by symbols. Each symbol can be decomposed into other symbols, recursively. Thus, for example, you might segment the entire instrumental piece down to the single note, or identify the individual sample in the case of a digitized piece.

Consider two musical structures: *Canzone* and Sonata Form. Fig. 1 shows the ancient form of *Canzone*, whose structure is A B A' where A' is a variation of A. One of the many forms that share this sequence of symbols is the *Aria col 'da capo'*, form that triumphed in the early 18th century.

Generally, A and B in this form contain tunes in different keys generated by different musical materials. A breakdown of these symbols is meant to split the melodies. Fig. 2 shows the Sonata Form, which according to the *New Grove Dictionary of Music and Musician* "is the most important principle of the musical form, or formal type, from Classical period well into the 20th century".

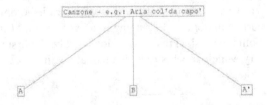

Fig. 1 Structure of Canzone

---

[1] *Musica est exercitium arithmeticae occultum nescientis se numerare animi.*

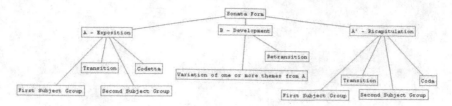

Fig. 2 Structure of Sonata Form

It can be noted that the first two levels of the graphs shown in Fig. 1 and Fig. 2 are identical. However, the second level in Fig. 2 is further split. In A and A' of the second graph there are two groups of Subjects, plus other items such as Transition and Codetta or Coda [15]. In B the materials of A are subjected to a greater degree of tonal, harmonic, and rhythmic instability than in A.

It is worth noting that *Canzone* and Sonata Form differ in their amount of information. In fact, in the Sonata Form sections of type A contain more information than the homologous structures of the *Canzone*. Furthermore, in Sonata Form B differs from A: B is not just different from A in the content of the materials as they are in *Canzone*, but because B is a process of variation of the materials of A with a larger quantity of information.

In the hypthesis of a phylogenetic relationship between the two forms, B of Sonata Form is the youngest structure in terms of the theory of evolution.

3 Trees in Music

Rens Bod, an expert of Computational Linguistics, wrote: "It is widely accepted that the human cognitive system tends to organize perceptual information into hierarchical descriptions that can be conveniently represented by tree structures" [2]. In fact, trees have been used to describe the linguistic perception [4], the music perception [8] and the visual perception [11]. An interesting unification of these perspectives is proposed in "A Unified Model of Structural Organization in Language and Music" [2].

The analysis discussed in the previous section illustrates that in practice the music piece can be represented by a tree structure in which each node is identified by a symbol.

In this work, the tree that produces the music piece is built using generative grammar [13].

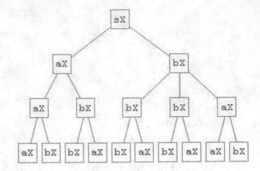

Fig. 3 Tree generated by grammar G

3.1 Formal Grammar

A grammar G is a quadruple (X, V, S, P) where:

- X is the alphabet of terminal symbols;
- V is the alphabet of nonterminal symbols;
- S is the start symbol (which belongs to V);
- P is the set of pairs (v, w) of symbols called production or rewriting rules. They are built on the union of the two alphabets, X and V, and are denoted as A → B.

S is the start symbol to which rules are applied, then rules are applied to the results previously obtained.

For example, let grammar G be formed by the quadruple (X, V, S, P):

X = {sX, aX, bX};

V = {sV, aV, bV};

S = sV;

initialRules = {sV → sX[aV, bV] , aV → aX[aV, bV] , bV → bX [bV,bV, aV]};

endingRules = {aV → aX[aX , bX] , bV → bX[bX , aX]}.

Applying the initialRules twice and the endingRules once results in the tree shown in Fig. 3 results.

3.2 Tree nodes

In the second section, it was shown that the tree symbols of a musical analysis contain the music piece or parts of it. Instead, in a tree structure used for composition the nodes contain mechanisms for generating parts of the musical discourse.

The node can be:

- non-terminal or terminal, i.e. leaf of the tree;
- non-composer or composer, i.e. generating music.

All tree nodes contain:

- one Symbol. One property of the Symbols is: parameter fields *pf1*, control signals *ks1,* sound gestures *g1* and algorithms *a1* are associated to the symbol *sym1*, regardless of which node *n* contains *sym1*. Type of the node is not associated to *sym1*;
- one Type of the node (which is addressed in Section 3.3);
- the values of zero or more Parameter Fields for the computation of the global structures, i.e. structures common to all nodes, such as metrical or harmonic structure of the piece;
- some Control Signals for the Algorithms for the Gesture Generation or Variation. The Control Signals are used to create Predictability, Surprise and Tension as proposed in [6].

The non-terminal nodes also contain the rewriting rule, which has been calculated or chosen, for the generation of the nodes of the next level. The composer nodes can contain the following items:

- zero or more Sound Gestures, described in a matrix form;
- zero o more Algorithms for Gesture Generation [5], [15];
- zero o more Algorithms for Reading the values of the Parameter Fields, the Control Signals or Gestures contained in the other nodes. The nodes which are likely read are:
  - nodes that lie on the same branch;
  - nodes that lie on the same level;
- zero or more Algorithms of Gesture Variation.

It is worth noting that the algorithms that use the data which lie along the same branch, partial or complete, facilitate the perception of hierarchy and the overall vision that controls the materials generated. Moreover, the algorithms that use the data which lie on the same level simulate the composer's analytic listening.

3.3 Node Types

All nodes of a given level follow one after the other over time. In music, however, it is sometimes necessary to have nodes at the same level which proceed in parallel, i.e. nodes can generate two or more Gestures of the same musical quality as, for example, two or more subjects superimposed.

The difference between the two Types of node can be shown by an example. Let the composer nodes *s* and *p* be one node of Series Type and Parallel Type, respectively. Let us suppose that the productions of their Gestures, i.e. of their music, are timed between the instants [*ts0, ts1*] and [*tp0, tp1*]. Let us also suppose that *s* and *p* are both fathers of two composer nodes. The timing of the materials produced by the children of *s* will be a partition of the interval [*ts0, ts1*], whereas the timing of both the children of *p* will be given by the whole interval [*tp0, tp1*].

The use of this formalism makes it possible to generate multidimensional trees. The multidimensionality of the trees permits the representation of the multimedia work.

To conclude, the model proposed supports both music and multimedia composition. As examples, Fig. 4, 5, 6 show frames of audio-video works by Sin[x]Thésis, a group for Multimedia Research and Production that I founded and directed.

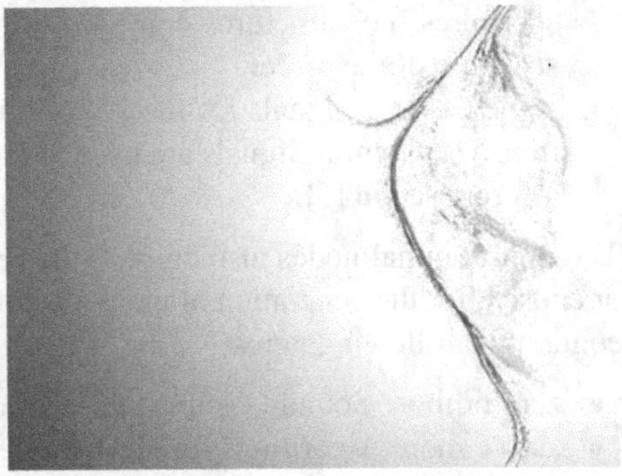

Fig. 4 Frame of audio-video opera: *Una Semplice Verità* by Giuseppe Salatino

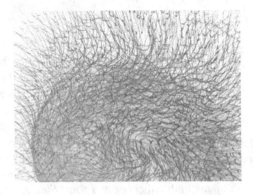

Fig. 5 Frame of audio-video opera: *Fantômes Électrique* by Antonio Mazzotti

Fig. 6 Frame of audio-video opera: *Studio sull' Intonazione della Carne* by Francesco Abbrescia

4 Conclusions

The model which I have been designing over the past years has been adopted as a tool by a small community of artists who have been investigating the deep connection between sound and emotional meaning. This bond is a fundamental part of the composer's research; it is also the most difficult part of our work to explain.

Models can produce highly complex musical objects. However, only the objects selected by the composer will constitute the new language used for expressing himself. Models are meaningless without the composer's knowledge. Without his profound knowledge, music cannot exist.

5 Acknowledgments

The author wishes to thank Anna Balenzano, his students and the artists of Sin[x]Thésis for the exciting and challenging project which they have been carrying out.

References

1. G. Assayag, H. G. Feichtinger, J.F. Rodrigues, Mathematics and Music. Springer, Berlin, 2002.
2. R. Bod, A Unified Model of Structural Organization in Language and Music. Journal of Artificial Intelligence Research 17: 289-308, 2002.
3. E. Campbell, Boulez, Music and Philosophy. New York, Cambridge University Press, 2010.
4. N. Chomsky, Aspects of the Theory of Syntax. Cambridge, MIT Press, 1965.
5. C. Dodge, T. A. Jerse, Computer Music. New York, Schirmer 1985.
6. D. Huron, Sweet Anticipation. Cambridge, MIT Press 2006.
7. V. Jankélévitch, La Musique et l'Ineffable. Paris, Seuil, 1961.
8. F. Lerdahl, R. Jackendoff, A Generative Theory of Tonal Music. Cambridge, MIT Press, 1983.
9. G. Ligeti, Wandlungen der musikalischen Form. Die Reihe 7: 5-17, 1960.
10. G. Mazzola, The Topos of Music. Basel, Birkhäuser, 2002.
11. D. Marr, Vision. San Francisco: Freeman, 1982.
12. New Grove Dictionary of Music and Musician, Macmillan, London, 2001, s.v.Sonata Form
13. G. Nierhaus, Algorithmic Composition. Wien, Springer, 2009.
14. L. Nono, Luigi Nono candidato del PCI con i lavoratori (1963). In: Scritti e colloqui, vol. 1, Angela Ida De Benedictis and Veniro Rizzardi (eds.). Lucca, LIM, 2001.
15. C. Rosen, Sonata Forms. New York, Norton, 1980.
16. I. Xenakis, Formalized Music. Revised edition. New York, Pendragon Press, 1992.

Printed in the United States
By Bookmasters